POPULATION MATTERS

Tony Champion is Senior Lecturer in Geography at the University of Newcastle upon Tyne. His principal research interests lie in the monitoring and analysis of population and social change in Britain, particularly their regional and local dimensions. He is chair of the IBG Population Geography Study Group and Council member of the British Society for Population Studies. Recent publications include *Counterurbanization* (Arnold, 1989), *Contemporary Britain: A Geographical Perspective* (Arnold, 1990), *People in the Countryside* (Paul Chapman, 1991) and *Migration Processes and Patterns: Research Progress and Prospects* (Belhaven, 1992).

POPULATION MATTERS

THE LOCAL DIMENSION

Edited by
Tony Champion

P·C·P
Paul Chapman
Publishing Ltd

Paul Chapman Publishing Ltd
144 Liverpool Road
London
N1 1LA

British Library Cataloguing in Publication Data

Population Matters: Local Dimension
   I. Champion, Tony
   304.6

ISBN 1 85396 2015

Typeset by Setrite Typesetters, Hong Kong
Printed and bound by Athenaeum Press Ltd, Newcastle-upon-Tyne.

A   B   C   D   E   F   G   H   9   8   7   6   5   4   3   2

CONTENTS

List of Contributors   vii
Preface   xi
List of Acronyms   xv

1   **Introduction:** key population developments and their local impacts   1
    Tony Champion

2   **Population pyramids and shifting sands:** targeting future
    investments to give the changing British public their just deserts   22
    Keith Dugmore

3   **Migration, places and quality of life:** voting with their feet?   33
    Allan Findlay and Robert Rogerson

4   **The demographic component of local government finance:**
    impacts on resources, needs and budgets   50
    Robert Bennett and Günther Krebs

5   **Population change and education:** schools rolls and rationalisation
    before and after the 1988 Education Reform Act   64
    Michael Bradford

6   **Fall-out from the demographic time-bomb:** a spatial perspective
    on the labour force effects of the 'baby-bust'   83
    Anne Green and David Owen

7   **Demography and house-building needs:** a critique of the
    'demographic bulldozer' scenario   101
    Dave King

8   **Going into a home:** where can an elderly person choose?   119
    Anne Corden and Ken Wright

9   **Healthy indications?** Applications of Census data in health
    care planning   136
    John Mohan

10 **Making waves?** The contribution of ethnic minorities to local
    demography   150
    Vaughan Robinson

    References   171
    Index   185

LIST OF CONTRIBUTORS

Robert Bennett is Professor of Geography at the London School of Economics. He has written extensively on public finance and local government, including work on Rate Support Grant, non-domestic rates/property tax, and local income tax. Publications include *Central Grants to Local Government* (Cambridge University Press, 1982), *Local Business Taxes in Britain and Germany* (Nomos, 1988) and *Local Economic Development in Britain and Germany* (Francis Pinter, 1991). He is currently working on local economic development, labour training policies, housing and business investment strategies.

Michael Bradford is a Senior Lecturer in the Department of Geography at the University of Manchester, where he is a member of the SPA (Spatial Policy Analysis) research group. Educated at St Catharine's College, Cambridge, he carried out postgraduate research at the University of Wisconsin in Madison and at the University of Cambridge. He is currently working on various aspects of the geography of education, an evaluation of the overall impact of inner city policy and an urban deprivation index from the 1991 Census.

Anne Corden is Research Fellow in the Social Policy Research Unit at the University of York. Her interests lie mainly in the field of social security and her research has focused on income-related benefits, including financial support for families with children and elderly people.

Keith Dugmore is Divisional Director with MVA Systematica, where he is responsible for geodemographic services. He began his census career twenty years ago with the then Greater London Council. Whilst at LAMSAC he managed the 1981 Census SASPAC project. He later spent five years at CACI where he set up its Public Services Group and was involved in acquiring and developing new data sets including population projections.

Allan Findlay is Senior Lecturer in Geography and research co-ordinator of the Applied Population Research Unit, University of Glasgow. His main research interests are in population geography and in particular in skilled international migration, rural–urban migration and refugee flows. He has also worked as a Research Associate of the United Nations International Labour Office.

Anne Green is Senior Research Fellow at the Institute for Employment Research, University of Warwick, where she contributes to the Institute's Local Development and Human Resources Initiative. Her main research interests include local labour market analysis, migration, the development of local economic indicators and spatial dimensions of employment, unemployment and demographic change. She has recently undertaken research projects on long-term unemployment and women returners.

Dave King is Professor of Strategic Housing Studies at Anglia Polytechnic University, Chelmsford, where he is Head of the Division of Housing, Rural Development and Planning and Head of the Population and Housing Research Group. He is the designer of the Chelmer Model and acts as demographic consultant to a number of public and private sector organisations, including the Housebuilders Federation. He is on the executive group of the BURISA (British Urban and Regional Information Systems Association) and is UK representative on the pan-European Urban Data Management Society.

Günter Krebs studied geography and economics in Giessen and Munich. He has been a lecturer at the Free University of Berlin and has held research posts in Berlin, in Cambridge and at the London School of Economics. His main research interest lies in regional and urban economics and he has published on a wide range of topics, including urban property markets, regional disparities, local government finance and local economic development.

John Mohan is Lecturer in Geography at Queen Mary and Westfield College, London; he has also held an ESRC Postdoctoral Fellowship (1986–9) and a Harkness Fellowship (1992–3) while based there. He has worked on geographical aspects of health care provision and health inequalities for some twelve years, publishing numerous book chapters and journal articles as well as carrying out consultancies for health authorities; he co-authored *Mapping the Epidemic* (Health Education Authority, 1990) and *Commercial Medicine in London* (Greater London Council, 1985) and is presently finalising a book on *Health Care Policy in Britain: from the Royal Commission to the White Paper*.

David Owen is Research Fellow at the Centre for Research in Ethnic Relations, University of Warwick. He is currently responsible for the development of a National Ethnic Minority Data Archive, bringing together and analysing secondary data sources such as the Census of Population, the Labour Force Survey and the General Household Survey, which contain information on the ethnic minority population. He has long-standing research interests in the handling of large data sets, spatial data processing techniques and local labour market analysis.

Vaughan Robinson is Senior Lecturer in the Department of Geography at University College Swansea, where he is also a Director of the Migration Unit. His research interests are in ethnic minorities and forced migration. He has authored or (co-)edited *Transients, Settlers and Refugees* (Clarendon, 1986), *The International Refugee Crisis: British and Canadian Responses* (Macmillan, 1992) and *Geography and Refugees: Patterns and Processes of Change* (Belhaven, 1993).

Robert Rogerson is Lecturer in Geography at the University of Strathclyde and Director of the Quality of Life Group. His current research interests lie in urban social studies and quality of life research. He is presently engaged in studies of quality of life and its role in population redistribution in urban and rural Britain and research on place marketing and residence selection.

Ken Wright is Deputy Director of the Centre for Health Economics at the University of York. His main research interests concern community care and his recent research into the care of elderly people has concentrated on the development of alternative patterns of domiciliary, residential and nursing home care.

PREFACE

The first chapter of Heather Joshi's edited collection *The Changing Population of Britain* ends with the statement 'Population does indeed matter' (Hobcraft and Joshi, 1989, p. 11). The same message comes across clearly in John Ermisch's earlier path-setting monograph entitled *The Political Economy of Demographic Change* (Ermisch, 1983) and his more recent *Fewer Babies, Longer Lives* (Ermisch, 1990); and it must have provided the inspiration which induced David Coleman and John Salt to produce their massive tome *The British Population* (Coleman and Salt, 1992). It is also the impression conveyed on an almost daily basis by the coverage of demographic, social and related topics in the news media. Matters covered most frequently include the pressure placed on social and medical services by the increasing numbers of the very elderly, the implications of lower birth rates for labour supply and supporting pension payments, the problems of adjusting school provision to match the declining number of children, and the various issues raised by the rapid growth of the ethnic minority population. This degree of interest reflects the major developments currently occurring in population composition and demographic behaviour, as evidenced most notably by fundamental changes in the role of women and the nature of the family and as highlighted by terminology such as the 'second demographic transition' and the 'demographic revolution'.

The justification for this book is that there is a growing recognition of the need to go beyond this broad level and examine the spatial dimensions and local impacts of these developments. This relatively new-found interest in the local dimension is reflected in the substantial use made of the 1981 Census's Small Area Statistics (much greater than after the 1971 Census) and in the strong turnout for meetings dealing with the implications of local population trends for business and government (the theme of a joint meeting of the British Society for Population Studies and the Population Geography Study Group in 1988). Research demonstrates that localities vary markedly in their population composition in terms of age structure, household types, ethnic composition and other socio-demographic characteristics, accounting for the amount of effort which both public and private sectors put into improved targeting of client groups. Studies also show how rapidly the profiles of particular places can change, placing a great premium on up-to-date intelligence and careful forecasting.

Up to now, however, most of the attention in the academic literature has been focused on the essentially demographic aspects of these changes and on their implications for central government and for society as a whole. This point was brought home to me very forcefully when three years ago, stimulated by the Joshi and Ermisch books, I decided – along with Social Policy lecturer

Peter Selman – to develop a third-year undergraduate course for Social and Environmental Science students at the University of Newcastle upon Tyne. At that time, there was available only one book that looked in some depth at the spatial dimension, namely *Planning for Population Change* (Gould and Lawton, 1986), which spanned the Third World as well as the European scene and concentrated on the four topics of labour supply, social provision, health care and education. There was clearly a case for a collection which had the same aim of linking demographic changes with their policy implications but which covered a wider range of topics for just the British context. This is the central purpose of *Population Matters: The Local Dimension*.

There are at least three components to this task. First and most basically, the book seeks to demonstrate that there are substantial geographical variations in population characteristics and demographic trends, underlining the fact that individual places are by no means faithful microcosms of the national population. Secondly, it highlights the main policy implications arising from local demographic profiles and trends, focusing primarily on those decision areas where local considerations are paramount or where there exists a local-scale mechanism for policy delivery. Thirdly, the book examines the extent and nature of the responses made by policy makers and planners to these challenges.

In addition, the book has a subsidiary aim which helps to explain the choice of timing. This is to provide an indication of the wealth of research opportunities which exist in this still relatively neglected area of population studies. The possibilities are increasing year by year, as new computing technology makes ever easier the manipulation of large data sets needed for local population analysis. This is true not just for big organisations like government departments and universities but also for private individuals working at home and students engaged in project work (as required on our third-year course).

The single most important data source for the study of local populations is the Population Census, for only this provides consistent nationwide data on a wide range of variables at a detailed geographical scale. By being prepared at the present time, this book is not only able to draw on the first results of the most recent Census, held in April 1991, but – more importantly – is designed to stimulate use of the Census data for local population analysis. The production of the machine-readable local statistics, which comprise a much larger set of tables than in 1981, is now scheduled to be completed by Spring 1993, around the time when this book is published, while the other data sets and topic volumes from the 1991 Census should all be available by the middle of 1994.

In organisation, the book starts with more demographic aspects and then moves on to policy themes. The first chapter develops the case for the book in more detail and, along with the next two chapters, provides plenty of evidence concerning the scale of place-to-place variation in population profiles, the speed with which local populations can change, and the processes underlying these patterns. The remaining seven chapters each deal with specific areas of policy concern, examining the details and significance of the underlying demographics and exploring the issues and policy responses. All the chapters involve some degree of population analysis, which can be used as a stimulus and guide to follow-up work with the latest data as it comes available; where relevant, authors have suggested ways in which the detailed results of the 1991 Census can be used for this work.

In being able to bring this book into being, I owe debts to many over a long period. My interest in developing this topic dates back to the publication of Ermisch's 1983 book and to Richard Lawton's Presidential Address (Section E) at the 1983 British Association Annual Conference on the theme of 'Planning for People'. The idea remained merely a 'gleam in my eye' for seven years; to adopt a population-related metaphor, actual conception did not take place until, thanks to the support of George Gordon and Tony Gatrell (respectively President and Recorder of the British Association's Section E at the time), I was invited to convene a one-day session on 'key population developments and their local implications' at the BA's 1991 Annual Conference, held at Plymouth. From this flowed the plan for this book. Half the papers derive from that meeting, albeit in revised form, while the others were specially commissioned subsequently. I am extremely grateful to all the contributors for being prepared to participate in the gestation process and for agreeing to my various suggestions and amendments; indeed, my apologies if they feel that the eventual birth of the book has involved the equivalent of a Caesarian operation!

My thanks also go to those involved in the final stages of gestation: my wife, Marilyn, who prepared the prelims, composite bibliography and index, and managed to distract Katherine and Victoria while the manuscript was being finalised; and Marianne Lagrange and Catherine Dunkling of Paul Chapman Publishing for their encouragement and exhortations and for the efficiency and high standard of their production work. Finally, it should be noted that several authors have made their own acknowledgements at the end of their chapters.

A last word: to say that this book is not meant to provide the last word, but really only a beginning or indeed a birth. This is why there is no concluding chapter, merely some preliminary observations at the end of Chapter 1; a formal conclusion will be more appropriate in, say, five years' time. It is confidently believed that the next few years will see a continued upwelling of local population analysis. If this book is judged to have played a part in encouraging this work, then I will feel that the effort of producing it has been worthwhile.

Tony Champion
Newcastle upon Tyne
November 1992

LIST OF ACRONYMS

ACORN:	A Classification of Residential Neighbourhoods
BURISA:	British Urban and Regional Information Systems Association
DGH:	District General Hospital
DHA:	District Health Authority
DSS:	Department of Social Security
ED:	Enumeration District
GHS:	General Household Survey
GIS:	Geographical Information Systems
GMS:	Grant Maintained Status
GRO(S):	General Register Office (Scotland)
ILEA:	Inner London Education Authority
LEA:	Local Education Authority
LEC:	Local Enterprise Company
LMS:	Local Management of Schools
NHP:	Nottingham Health Profile
NIMBY:	Not In My Back Yard
NOMIS:	National Online Manpower Information System
OPCS:	Office of Population Censuses and Surveys
RAWP:	Resource Allocation Working Party
RHA:	Regional Health Authority
RSG:	Revenue Support Grant
SAS:	Small Area Statistics
SEA:	Standard Expenditure Assessment
SMRs:	Standardised Mortality Ratios
SPA:	Spatial Policy Analysis
SSA:	Standard Spending Assessment
TEC:	Training and Enterprise Council
TTWAs:	Travel-to-Work Areas
TVEI:	Technical and Vocational Education Initiative

1. INTRODUCTION:

Key population developments and their local impacts

Tony Champion

At first glance the recent history of population in most of the developed world conveys an almost static picture, with frequent mention of terms like 'zero population growth'. Yet, as outlined in the preface, the reality is very different. There are strong forces of demographic and social change at work in Britain, as elsewhere; in particular, declining fertility, population ageing, the breakdown of traditional family and household arrangements, the greater participation of women in the labour force and the growth of non-white ethnic populations, as well as the knock-on effects of the 1960s/1970s baby boom and bust. Moreover, these changes in demographic structure and behaviour have been accompanied by a massive redistribution of the population, notably the exodus from the largest cities to smaller settlements and more rural areas.

The restructuring of Britain's population and the national-scale implications of this process are now reasonably well documented, as reflected in the wealth of material brought together by Joshi (1989) and Coleman and Salt (1992). By contrast, the local dimension has been relatively neglected in the literature for, though migration has generated considerable research interest in recent years (for reviews, see Champion and Fielding, 1992; Stillwell, Rees and Boden, 1992), the impact of these key developments on local population profiles and the implications for local planning and policy making have not received the same level of attention as that accorded to the national issues. As Compton (1990, p. 462) puts it, 'It is all well and good focusing on temporal trends in the aggregate, but this should not be at the expense of spatial patterns, which, at a single point in time, may exhibit greater heterogeneity than do aggregate trends through time.' In fact, one of the most impressive changes in British demography over the past decade has been the growth of interest in local populations, even though this development is not yet widely reflected in population texts.

The primary aim of this book is to go some way towards redressing this balance. It contains studies which document the extent of regional and local variation in population trends and characteristics and which provide examples of the many policy issues which local population developments raise. In so doing, it brings together a great deal of information arising from recent research in this area, but it also indicates the vast scope for further inquiry and the unparalleled opportunities now available for this work, particularly those offered by the arrival of results from the 1991 Census.

This introductory chapter aims to provide background for these studies. It makes the general case for sub-national perspectives and presents evidence on the importance of the local dimension. Then it outlines the principal messages conveyed by the other chapters and makes some concluding observations about their policy significance and the value of improved intelligence for decision-making in these areas. First of all, however, it is essential to underline the importance of the developments currently affecting populations.

Population matters

Interest in demographic issues has experienced a major resurgence over the past few years, as reflected in the burgeoning literature. There should be no surprise about this. As Joshi (1989, p. vii) puts it, 'The study of population is vital. It is vital in that it is concerned with the vital events of birth and death and other momentous transitions in people's lives. It is vital for informing forecasting and decisions in both public and private sectors. It is also vital . . . in the sense that it is a very lively branch of social science in Britain.'

The collection of essays in Joshi's book provides ample evidence of the importance of demographic considerations for individuals, communities, business and government. Fluctuations in the birth rate immediately affect the need for maternity beds, postnatal care, crèches, and nursery facilities. Subsequent progress of baby boom and bust cohorts through the age structure causes major challenges for a wide range of age-specific areas, including school provision, labour supply, housing needs and eventually pensions and elderly care. Housing, income support and related needs can arise from changes in the role of women, the strength of family life, the socio-economic composition of the population and the size and distinctiveness of ethnic minorities. Both international and internal migration will affect spatial patterns of population distribution and the demographic profiles of individual localities.

The current salience of population matters can also be attributed to the fact that many of the changes which have been taking place in demographic behaviour and population structures over the past quarter of a century have proved rather different from expectations. Forty years ago demographers were, with some justification, thinking that populations in the developed world were settling down in terms of moves towards stationary, or even declining, levels and towards stable structures. After all, several of these countries (notably in north-west Europe and North America) appeared to have completed their passage through the demographic transition, while many others (including southern Europe and Japan) were moving swiftly in the same direction, with fertility rates starting to fall rapidly after decades of mortality reductions. Yet, within a few years, birth rates were rising sharply through much of the developed world and then — just as population projections were being revised upwards to cope with this — they fell even more dramatically, giving rise to the marked boom—bust imprint on cohort size which will be working its way through the age structure until the middle of the next century.

Even more fundamentally, the mid 1960s are now being seen by some as the start of a new era in demographic history, conceptualised as the 'second demographic transition' (van de Kaa, 1987) and the 'demographic revolution' (McLoughlin, 1990). The principal feature of this new demographic regime is the decline of fertility from somewhat above the 'replacement level' of 2.1 births per woman to well below it. This development has been associated with a fundamental shift in norms and attitudes, which van de Kaa (1987, pp. 5—7) has denoted as a switch from altruism to individualism. Whereas the first transition to low fertility was dominated by concerns for family and offspring, the second emphasises the rights and self-fulfilment of individuals, the desire for people to realise more of their own potential, and an increasing emphasis on equality of opportunity and freedom of choice. Alongside basic considerations concerning the economic cost of children, social and cultural changes play a crucial role in the move away from marriage and parenthood. In particular, a close link has been identified between the development of new lifestyles and the emancipation of women, suggesting that improvements in the societal position of women have led directly to the decline of fertility and the rise of non-traditional living arrangements (Hoffman-Nowotny and Fux, 1991; Hopflinger, 1991; McLoughlin, 1990).

The scale of these developments at national level is very impressive. In the UK the total fertility rate in 1991 was only 1.82, down from 2.83 in 1965. In some other European countries the fall has been even more dramatic, the rate almost halving between the mid 1960s and the late 1980s in some cases; for instance, down from 2.55 to 1.32 in Italy, 2.50 to 1.36 in West Germany, 2.68 to 1.43 in Austria and 2.61 to 1.50 in Denmark. In relation to family break-up, the number of divorces per 100 marriages in the UK reached 42 by the mid 1980s, up from barely 1 in 10 in 1965 and comparable with the traditionally high-divorce countries of Sweden (46) and Denmark (45).

Taken together, these changes have had major impacts on age structure and household composition. In the 30 years since 1960 the proportion of those aged 65 years and over in the UK has grown from 11.9 to 15.6 per cent, with particularly rapid growth for those aged 75 and over, up from 4.2 to 6.9 per cent by 1990. The number of children aged under 16 fell by almost one-fifth between 1971 and 1990, when they made up barely 20 per cent of total population compared to 26 per cent in the early 1970s. The ageing of the population, along with lower fertility, higher divorce rates and the tendency for young people to set up separate households before marriage, has led to a substantial shrinkage in average household size, down from 3.1 persons in 1960 to 2.5 now. This has occurred most notably through an increase in the proportion of one-person households, now comprising 26 per cent in Great Britain and rising — though still well short of the level in Sweden (already 33 per cent in 1980). The stage has now been reached where in many developed countries at least three out of five households contain no dependent children at all (the figure for Great Britain was 66 per cent in 1991) and where a significant proportion of family-type households contain only one parent (20 per cent in Britain in 1990).

Moreover, the idea of a new demographic regime based on changing lifestyles applies just as much to migration as to fertility and household arrangements. It is not just that the two are directly linked, in that developments like increased childlessness and household instability will be associated with changes in people's propensity to move (Champion, 1991). There is also increasing evidence of the growing importance of lifestyle and 'quality of life' factors in stimulating certain types of residential moves and influencing choice of destination. Most notable in this context is the attention given to residential preferences in the 'counterurbanisation' literature (see Champion, 1989a, Chapter 2). The idea of something beyond the demographic transition in relation to migration is found in Zelinsky's (1971) 'hypothesis of the mobility transition', where a distinction is drawn between the advanced and the future super-advanced society, each with its own characteristic pattern of population movement. This has been applied notably to developments over time in the migration of older people (see Warnes, 1992, on the elderly mobility transition).

These developments have already begun to pose extremely important challenges for policy makers and decision takers. Since the mid 1960s, there has been a substantial contraction in the need for maternity facilities and school places nationally. Reduced levels of labour force entries from education was causing staff recruitment problems before the end of the 1980s, prompting employers to induce more women to keep working during their family-raising stage and encouraging more of the others to return to work afterwards. The knock-on effects of this so-called 'demographic time-bomb' include the ageing of the workforce and an increase in the ratio of those of pensionable age to the working-age population. Similarly, the increasing incidence of relationship breakdown and household fission is producing serious consequences for the care and support of children and the elderly and infirm, as well as creating housing and other problems for the people involved.

The increasing importance of the local dimension

Given the fundamental nature of these population developments, it is perhaps not surprising that attention has tended to be focused, first and foremost, on the national-scale changes and on their implications for central government. Nevertheless, interest in the local dimension has increased enormously over the past decade or so. There are several reasons for this, which provide the rationale for this book. These include the fact that many organisations need to relate directly to their local clientele and have an increasing ability to do this via modern technology and more detailed data sources, but the basic factor concerns the great diversity of local populations and the way in which they change over time.

A large number of organisations nowadays make use of local population estimates and projections, primarily those which provide services directly to people. In a society which is consumer-orientated and largely sympathetic to the provision of welfare services, a considerable proportion of both private and public investment is geared to providing goods and services to people within fairly easy reach of their homes. A single national centre is appropriate only for the few services which involve delivery to all addresses in Britain (e.g. mail order), very occasional visits (e.g. specialised medical advice) or a particular type of service or activity with a client group that lives close to it and is not represented elsewhere in the country (e.g. certain elements of central government, the national media and business services).

An indication of the range of applications is given by Joshi and Diamond's (1990) survey of users of demographic projections in Britain and Australia. Most numerous among the main types of responding organisations were local government, health authorities, private firms, central government, market-research consultancies and education authorities. In the British sample, the most commonly mentioned applications (besides the rather general heading 'research') were welfare provision, planning, staff recruitment, education, housing, transport and marketing, in that order. Moreover, in the present context, it is worth noting that, while users in the two countries were found to be generally satisfied with official projections, the need for better local-scale projections was a recurrent theme, as also was the call for better and more up-to-date information on the present situation. This point is developed more fully by Dugmore in Chapter 2 of this book.

This degree of interest in local population intelligence can be attributed both to need and to opportunity. In relation to the latter, the ability to study local populations in detail has been vastly increased by the appearance of machine-readable Small Area Statistics. First produced in Britain from the 1971 Census, the Small Area Statistics (SAS) were expanded in the 1981 Census and extended even further in 1991 to include a second major data set: the Local Base Statistics, containing a very much wider set of cross-tabulations than anything previously released, albeit for ward level rather than for the individual Enumeration Districts of SAS.

The growing need for local intelligence arises from the drive to greater efficiency, which has been a feature of both public and private sectors over the past few years. Under pressure from diminishing resources and central government exhortations, much greater emphasis is being put on the careful targeting of policy measures on the places with the most severe problems or the greatest prospects for recovery, as evidenced by successive reviews of regional policy and by the initiatives aimed at rejuvenating inner city areas. Similarly, the private sector has developed sophisticated targeting procedures for advertising, marketing and product monitoring.

None of this information would, of course, be required if it was not for the diversity of local populations. The most important and obvious reason for studying the local dimension, following up Compton's point above, is that places differ. Regional and local populations do not by any means constitute faithful microcosms of the national population. This is true both in terms of the rates of change in their size and in terms of their composition by age, gender, household type, socio-economic characteristics and so on. Not only does migration continue to produce considerable variations between places in rates of population change, but mortality also exhibits substantial spatial differentiation and so too does fertility, though to a diminishing extent.

Nor do places change over time in their characteristics and demographic behaviour in the way that would be expected from the national statistics alone. One reason for this is that the various key developments outlined in the previous section will have different impacts on places as a result of operating on the distinctive demographic structures which each locality has inherited from its past histories of fertility, mortality and migration. For instance, a nationwide rise in fertility rate would produce a much larger increase in births in a recently developed new town with an above-average proportion of couples than in a retirement area with a population of a similar size. A second reason is that migration not only leads to changes in total population numbers but can also alter the population composition of an area. This can be the case even where net migration is generating very little overall change in population size, if the characteristics of newcomers to an area are very different from those of the out-migrants.

A final point is that, with the decline in rates of natural increase, migration has assumed a much more important role than twenty years ago in determining whether places grow or decline in population size. The fall in birth rate has meant that population redistribution has virtually become a 'zero-sum game' whereby any increase in one place can take place only at the expense of another. Gone are the days when even substantial rates of net migration loss from the larger cities to other parts of Britain were offset by relatively high rates of natural increase. With the switch in emphasis from accommodating national growth, as in the 1960s, to the present reality of redistributing an essentially fixed number of people come the problems of spare capacity and loss of investment confidence in the adversely affected areas.

In sum, the local dimension has taken on much greater importance in population matters in recent years. This is partly due to an autonomous growth of interest by service-providing agencies, arising from their increasing ability to obtain information accurately and quickly. But it also derives from the fact that the local population scene is constantly changing, being even more unstable than the national population which itself is undergoing a major transformation. Migration means that relatively few places are changing in overall population size at anything close to the overall national rate. Meanwhile, national changes in demographic behaviour and population structure will be reflected in most places round the country, but their precise impact will vary spatially according to the inherited characteristics of individual places. Past demographic processes have produced a very varied mosaic of local populations, and these cannot be expected to evolve in a consistent manner in the future. The next section demonstrates the scale and nature of these regional and local differences.

A spatial perspective on the British population

This section concentrates on two aspects of the local dimension: the changing size of regional and local populations and differences between places in population characteristics. The aim is not to be comprehensive, but rather to provide examples of the scale of variations and to give some insight into their nature, drawing particularly on those relevant as background to the studies contained in this book. More detailed accounts of these patterns can be found in Coleman and Salt (1992), as well as in a population geography text like Jones (1990) and books on the geography of economic and social change in Britain (e.g. Champion and Townsend, 1990).

Regional and local population numbers

Table 1.1 provides the most up-to-date information available at the time of writing (October 1992) on population distribution and change for the standard regions – the most common level of presenting sub-national patterns. The 1991 data are the provisional mid-year estimates produced in the light of the 1991 Census results and will be replaced by final figures during 1993, but any adjustments are unlikely to undermine the following broad observations. The regions are ranked in the table according to their rate of growth over the previous ten years.

The first point to note from Table 1.1 is that past patterns of growth have led to a very uneven distribution of the British population. Most impressive in absolute terms is that almost one-third lives in the South East, which accounts for less than one-eighth of Britain's land area, giving a density of well over twice the national average. The North West, however, is the most heavily populated region, with 868 persons per km² – a density which is 13 times higher than that of the lowest density region, Scotland with only 66. This is probably the single most important spatial differential in terms of its effect on providing for people, and becomes even more significant at more local scales; for example, the range of population density at the next administrative tier was (in 1991) from 4308 persons per km² in London to only 8 in the Highland region of Scotland.

Similarly, in relation to population trends over time, Table 1.1 shows that, even at this broad regional scale, there is a wide variation in change rates

Table 1.1 Population distribution and change in Great Britain, 1981–91, by region

Region	1981 000s	1991 000s	Density persons/km²	1981–91 change 000s	%
East Anglia	1,894.7	2,091.1	166	196.5	10.4
South West	4,381.4	4,723.4	198	342.0	7.8
East Midlands	3,852.6	4,025.7	258	173.1	4.5
South East	17,010.6	17,557.6	645	547.0	3.2
Wales	2,813.5	2,886.4	139	72.9	2.6
West Midlands	5,186.6	5,254.8	404	68.2	1.3
Yorkshire & Humberside	4,918.4	4,954.2	321	35.7	0.7
North	3,117.4	3,084.2	200	−33.2	−1.1
North West	6,459.2	6,377.4	868	−81.7	−1.3
Scotland	5,180.2	5,100.0	66	−80.2	−1.5
Great Britain	54,814.6	56,054.8	245	1,240.3	2.3

Note: Data relate to estimates of resident population at mid-year. Density refers to 1991. 1991 data are provisional. Regions are ranked in order of change rate.
Source: Office of Population Censuses and Surveys and Registrar General Scotland. Crown Copyright.

around the national figure of 2.3 per cent for the decade. Three regions are estimated to have lost population by at least 1 per cent, while one region grew by some 10 per cent. The regional pattern reveals the importance of the North–South divide, as well as the influence of the urban–rural shift. It is the three northernmost regions which have been less dynamic in demographic terms, with Yorkshire & Humberside and the West Midlands also lying below the national rate. Wales comes next, while top of the list are the four regions which, since the early 1980s, have been recognised as the full extent of the 'South' (i.e. excluding the West Midlands). The urban–rural shift is particularly reflected in the much stronger growth of East Anglia and the South West compared to the South East, the latter being adversely affected by possessing London (still six times the size of Britain's next largest city) and by the pressures which this has imposed on the shire counties in that region.

Disaggregation of these population changes into two broad components reveals the basis of this regional variation. As shown in Table 1.2, while natural change (the surplus of births over deaths) is a more important contributor to population growth at national level than is migration, the latter is by far the greater factor in producing regional differences in growth rate. The percentage point range between the highest and lowest rates of migratory change (11.45 points between −2.62 and +8.83) is four times the range for natural increase (2.81 points). For this reason, the patterns of migratory change shown in Table 1.2 broadly mirror those of overall change shown in Table 1.1. They emphasise the North–South distinction, placing the West Midlands firmly in the 'North' and showing that nowadays Wales is clearly part of the 'South' in terms of migration attractiveness. Table 1.2 also confirms the strong performance of East Anglia and the South West, benefitting from both the North–South drift and the urban–rural shift. The urban–rural factor works strongly against the South East, struggling to match the national level of migratory change, while the North West suffers from the dual disadvantage of northern location and high population density.

The patterns of natural change differ considerably from this pattern (Table

Table 1.2 Components of population change in Great Britain, 1981–91, by region

Region	Natural change		Migration and other changes	
	000s	%	000s	%
East Anglia	29.2	1.54	167.3	8.83
South West	−6.1	−0.14	348.1	7.94
East Midlands	76.4	1.98	96.7	2.51
South East	451.6	2.65	95.5	0.56
Wales	25.0	0.89	48.0	1.71
West Midlands	138.3	2.67	−70.1	−1.35
Yorkshire & Humberside	65.1	1.32	−29.4	−0.60
North	18.1	0.58	−51.3	−1.64
North West	87.1	1.36	−169.4	−2.62
Scotland	27.0	0.52	−107.2	−2.07
Great Britain	911.7	1.66	328.1	0.60

Note/source: as for Table 1.1.

1.2). Indeed, it is one of the two fastest growing regions which exhibits the lowest rate of natural change: the South West, the only region with a surplus of deaths over births between 1981 and 1991. This is a long-term characteristic of this region, arising from the fact that a combination of retirement in-migration and the net out-migration of young adults has led to relatively large numbers of the elderly and a relatively low proportion of people of family-building age. The most impressive feature of Table 1.2, however, is the major contribution of the South East, which is responsible for virtually half of the nation's population growth due to natural increase. This is not just because of its large population base, but also because during the decade this region has recorded a well above-average rate of natural increase, second only to the West Midlands.

Clearly, regions vary not only in their overall rates of population growth, but also in the relative importance of the demographic components directly responsible. The South West provides the classic example of migration-based growth, whereas the South East has been depending almost entirely on natural increase to sustain its population. The West Midlands has been relying on its strong natural growth to offset its sizeable migration losses, but the three northernmost regions suffer from the double problem of low natural change and substantial net out-migration. It is worth noting, however, that even had the North West and Scotland achieved the national average rate of natural change, they would still have experienced depopulation over the decade: overall national growth is now at such a low level that gains in one area are very likely to lead to losses in another − a long way from the demographically dynamic days of the 1960s, where even relatively high levels of migratory loss could be offset by natural increase. No longer can the present population level be seen as the guaranteed absolute minimum to be found there at any future time.

This point about the 'zero-sum game' nature of current population redistribution carries even more force at more local scales. Figure 1.1 shows that population losses are relatively widespread at the level of counties and Scottish regions. Between 1981 and 1991 particularly high rates of loss were recorded by areas with the largest northern conurbations, notably Merseyside and Strathclyde. At the other extreme, there is a zone of strong growth stretching from Cornwall to Lincolnshire and Norfolk, with rates reaching 10.5 per cent or more in Buckinghamshire, Cambridgeshire, Cornwall and Dorset and also a compact area of growth comprising Hereford & Worcester, Powys and Shropshire.

As the scale of analysis drops from regional to more local scales, the importance of the urban−rural factor in growth differentials clearly becomes greater. This point is demonstrated further by the district-level analysis of 1981−91 trends shown in Table 1.3. Metropolitan Britain, comprising London, the English metropolitan counties· and the Central Clydeside Conurbation, are estimated to have 'lost' over a third of a million people over the decade, a decline of 1.8 per cent which is 4 percentage points adrift from the national rate. Meanwhile, the remainder of the country saw its population grow by 1.6 million, an increase of 4.8 per cent or twice the national rate. The urban−rural distinction is also found within non-metropolitan Britain,

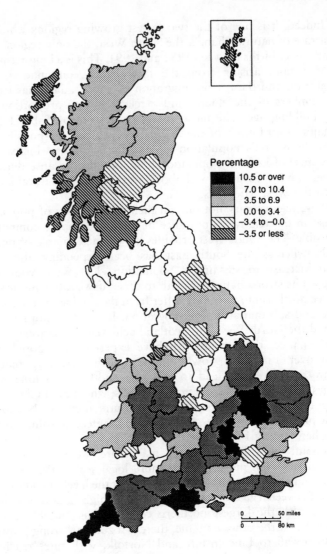

Fig. 1.1   Population change, 1981–91
           (*Source*: mid-year population estimates from the Office of Population
           Censuses and Surveys and Registrar General Scotland).

with a growth-rate progression from −1.2 per cent for large cities through
+1.8 for small cities and 5.7 for 'urban and mixed urban–rural' districts to
+7.6 per cent for 'remoter, mainly rural' districts.

The influence of other factors can also be seen at this more detailed scale,
albeit operating with the context of this general urban–rural gradient. The
North–South dimension is evident in the greater declines sustained by the
metropolitan districts than by London, where it is interesting to note that the
inner boroughs − after decades of poorer performance than outer London −

Table 1.3 Population change in Great Britain, 1981–91, by type of district

Type of district	1981 000s	1991 000s	Change 000s	%
Great Britain	54,814.5	56,054.8	+1,240.3	+2.3
Metropolitan Britain	19,831.6	19,471.9	−359.7	−1.8
London	6,805.6	6,803.1	−2.5	−0.0
Inner boroughs	2,550.1	2,566.4	+16.2	+0.6
Outer boroughs	4,255.4	4,236.7	−18.7	−0.4
Metropolitan districts	13,026.0	12,668.8	−357.2	−2.7
Principal cities	4,324.3	4,089.2	−235.1	−5.4
Other districts	8,701.7	8,579.6	−122.1	−1.4
Non-metropolitan Britain	34,982.9	36,582.9	+1,600.0	+4.6
Large cities	3,674.7	3,630.7	−44.0	−1.2
Small cities	1,922.9	1,958.2	+35.3	+1.8
Industrial areas	7,440.3	7,564.9	+124.6	+1.7
New towns	2,548.9	2,737.2	+188.3	+7.4
Resort, port and retirement	3,367.9	3,633.9	+266.0	+7.9
Urban and mixed urban/rural	9,840.3	10,400.8	+560.5	+5.7
Remoter, mainly rural	6,188.0	6,657.2	+469.2	+7.6

Note: see notes and sources for Table 1.1. The 1981 data are adjusted to take account of boundary changes between 1981 and 1991. 'Metropolitan' includes the Central Clydeside Conurbation area.

managed to gain people over this period, largely due to the growth of Tower Hamlets where much of the Docklands reconstruction has been focused. Economic structure, local environmental factors and planning policy have also played their part, notably in the below-average growth of industrial areas and in the high growth rates recorded by the resorts and retirement areas and by districts with new towns – these latter two types being best represented along the South Coast and in the Cornwall-to-Lincolnshire zone.

Finally, at the more disaggregated scale of the individual district, the range of variation is much wider again. According to the provisional estimates for mid 1991, some places saw their population level drop by more than 1 in 10 from 1981, notably Clydebank (−12.2 per cent) and Glasgow and Knowsley (Merseyside) (both −11.2), while others like Inverclyde (−9.5), Liverpool (−8.2) and Salford (−7.9) also experienced substantial contraction. The top ten districts are led by that symbol of 1980s growth, Milton Keynes, with an increase of 42 per cent over the decade, but districts growing by at least 20 per cent include Kincardine & Deeside (Grampian), Forest Heath (Suffolk), Wokingham (Berkshire), Gordon (Grampian) and Huntingdonshire.

Population profiles

Two aspects of population composition have been highlighted as being important in the study of the local dimension. One is that local populations will be participating in, or in another sense contributing to, the overall national changes in population structure, admittedly to varying degrees. The other is that, at a single point in time, places can vary markedly in their socio-demographic profiles. The latter will reflect places' past history of population change and will also influence the extent to which the places share in national trends. Moreover, when changes in population composition are related to the patterns of population redistribution described above, some very large changes can be found in the absolute size of individual population subgroups. Here this variability is explored with respect to three characteristics: age structure, limiting long-term illness and ethnic group.

As pointed out earlier, along with spatial shifts, the most important feature of population change in Britain over the past two or three decades has been the redistribution of people between age groups. This point is illustrated in Table 1.4, which reveals the marked reduction in numbers of children since 1971 and the major increases in numbers of 25–44 year olds and particularly of those aged 75 and over. The trends in age composition produced by these changes in numbers are so significant that few localities can have escaped their impact altogether. Certainly, aspects of changing age structure feature strongly in Chapters 5, 6 and 7 of this book, as the spotlight moves from school places to labour supply and then on to household formation.

On the other hand, the actual experience of each place will also be influenced by its overall growth rate, its inherited age structure and any special features of its net migratory exchanges. This point is exemplified by the degree of local variation in the changing numbers of people aged 75 and over, as shown in Table 1.5. Ten districts are estimated to have seen this group increase by more than 50 per cent during the 1980s, well over twice the national rate of 21 per cent, while several places recorded virtually no change, including two with reductions. The former set largely comprises new towns and relatively new suburban areas that were either growing rapidly in the 1980s or during that decade were experiencing the ageing of large numbers of previous younger newcomers, in both cases building on a relatively small base of this group. The slow-growth set is more varied, including established retirement towns where the numbers moving into this age group have barely kept pace with mortality losses as well as inner city areas and industrial districts which lose older people through out-migration and/or are gaining younger people through international immigration and the rapid

Table 1.4 Change in age structure and size of age groups, 1971–90

Age group	1971 000s	%	1990 000s	%	Change in numbers 000s	%
0–4	4,553	8.1	3,841	6.7	−712	−15.6
5–14	8,916	15.9	7,079	12.3	−1,837	−20.6
15–24	8,144	14.6	8,473	14.8	+329	+4.0
25–34	6,971	12.5	8,805	15.3	+1,834	+26.3
35–44	6,512	11.6	7,890	13.7	+1,378	+21.2
45–59	10,202	18.2	9,438	16.4	−764	−7.5
60–74	7,986	14.3	7,904	13.8	−82	−1.0
75+	2,644	4.7	3,980	6.9	+1,336	+50.5
United Kingdom	55,928	100.0	57,411	100.0	+1,483	+2.7

Source: calculated from Population Trends 66, Table 6, p. 59.

Table 1.5 Districts with highest and lowest percentage increases in numbers of residents aged 75 and over, 1981−90

District	Change in numbers	% 75+ 1991	District	Change in numbers	% 75+ 1991
Highest			Lowest		
Crawley	+95.5	4.5	Haringey	−5.3	5.5
Corby	+79.5	4.5	Hove	−3.7	13.4
Stevenage	+66.0	4.6	Kensington & Chelsea	+0.6	6.2
Northavon	+62.9	5.1	Colwyn	+1.4	12.0
Welwyn Hatfield	+59.1	6.8	Worthing	+1.4	15.1
Broxbourne	+58.8	5.1	Burnley	+2.7	7.0
Bracknell Forest	+54.8	4.5	Blackburn	+2.8	9.8
Milton Keynes	+53.1	4.2	Nairn	+3.1	8.1
Solihull	+53.0	5.8	Newham	+3.7	5.3
Three Rivers	+51.9	7.3	Hastings	+3.9	10.7

Source: Regional Trends 27, Table 15.1; 1991 Census Monitors, Table E.

natural increase of ethnic minority populations. Whatever the causes, however, the principal point here is the large scale of variation between places, with important implications for the rate of growth in the need for residential care homes to move into (the topic of Chapter 8) and in the level of funding for health care in general (see Chapter 9).

The broader context of age structure is shown in Figure 1.2. At county level the proportion of the population of pensionable age in 1991 ranged from 26.4 per cent in East Sussex and the Isle of Wight to 14.6 per cent in Buckinghamshire. In general, the highest proportions are found in more rural areas, notably those with coastal towns which have proved particularly attractive as destinations for retirement migration as well as having long traditions of out-migration by young adults − in contrast to the areas of northern Scotland affected by the in-migration of young workers in oil-related industries over the past twenty years. Besides the latter, the youngest populations occur in the areas of recent urbanisation, notably to the north and west of London.

An aspect related to age structure is the incidence of long-term illness, with the health-care and financial-support implications that this involves. Information produced by a new question in the Population Census shows that in some counties 1 in 5 people were suffering from long-term illnesses which limited their activities, whereas in others the level was less than half this. As shown in Figure 1.3, the lowest proportions occur in the counties north and west of London, corresponding to the youthful populations there. The high-incidence counties, however, do not closely match those with the largest proportions of older people shown in Figure 1.2, with, for instance, lower than expected incidence in some of the South Coast counties and higher than expected levels in south Wales, central Scotland and parts of northern England. Here again, various factors are likely to be involved, including work in unhealthy occupations (like coal-mining) and the fact that it is the healthiest older people who tend to migrate from metropolitan areas into retirement counties.

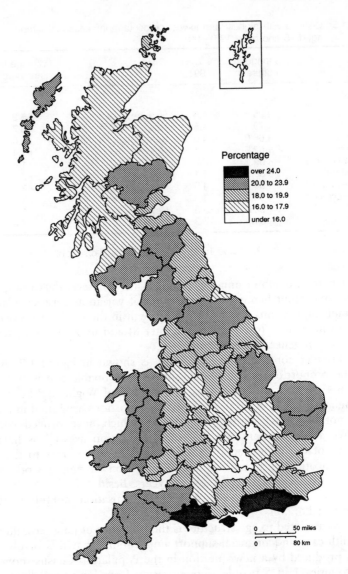

Percentage

- over 24.0
- 20.0 to 23.9
- 18.0 to 19.9
- 16.0 to 17.9
- under 16.0

0 _____ 50 miles

0 _____ 80 km

Fig. 1.2 Persons of pensionable age and over as a proportion of all residents, 1991
(*Source*: 1991 Census Monitors, Table E: Residents by age).

Finally in this set of examples designed to demonstrate the degree of spatial
diversity across Britain, Figure 1.4 shows the contribution made to local
demography by people of non-white ethnicity. As with long-term illness,
1991 was the first time that this information was collected by the Population
Census, being included in the questionnaire as a result of over ten years of
campaigning by service-providers because of the importance of the issues
raised by this group, as documented further in Chapter 10. This group of

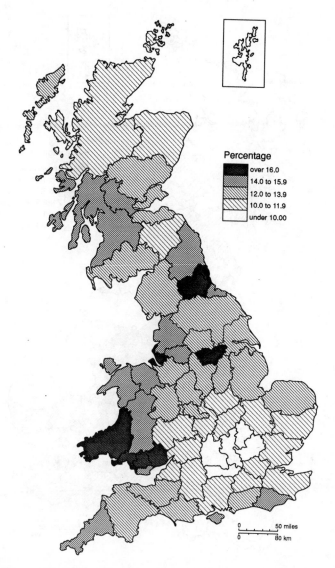

Percentage
- over 16.0
- 14.0 to 15.9
- 12.0 to 13.9
- 10.0 to 11.9
- under 10.00

```
0            50 miles
0            80 km
```

Fig. 1.3 Persons with limiting long-term illness as a proportion of all residents, 1991 (*Source*: 1991 Census Monitors, Table E: Residents by age).

people is characterised by an extremely distinctive distribution. London's position is outstanding, with 1 in 5 of its 6.8 million people classing themselves as non-white, four times the national average. West Midlands, Leicestershire, Bedfordshire and West Yorkshire are the next highest counties, in that order, and there is a further handful of counties with levels of over 4 per cent, but many parts of the country have virtually none (Figure 1.4).

There is also a strong degree of concentration within counties. For instance,

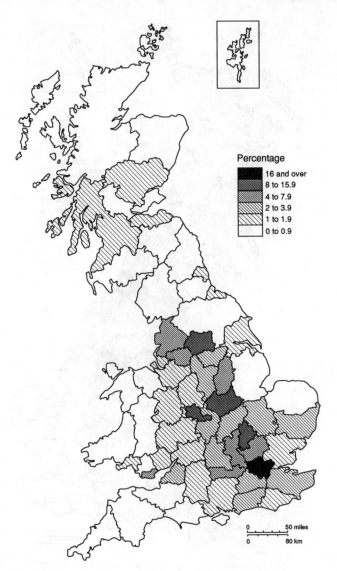

Fig. 1.4   Non-whites as a proportion of all residents, 1991
(*Source*: 1991 Census Monitors, Table J: Ethnic group of residents).

80 per cent of Leicestershire's non-white population lives in Leicester itself,
making up 28.5 per cent of the city's residents. The highest concentrations at
the district level, however, are found in London, where nearly half of Brent's
population is non-white and the proportion is over 30 per cent in several
other boroughs (Table 1.6). Furthermore, there is diversity in the patterning
of individual ethnic groups, as illustrated by the large proportions of blacks
in Hackney, Lambeth and Haringey, Indians in Leicester, Ealing and Harrow,
and Bangladeshis in Tower Hamlets.

Table 1.6 Districts with highest proportions of non-white residents, 1991

District	All non-white	Black	Indian	Pakistani	Bangladeshi	Other
Brent	44.8	16.5	17.2	3.0	0.3	7.9
Newham	42.3	14.4	13.0	5.9	3.8	5.2
Tower Hamlets	35.6	7.1	1.0	0.7	22.9	3.7
Hackney	33.6	22.0	3.5	1.0	1.8	5.4
Ealing	32.2	7.1	16.1	2.7	0.3	6.2
Lambeth	30.3	21.8	2.1	0.8	0.7	4.8
Haringey	29.0	17.1	3.6	0.7	1.5	6.1
Leicester	28.5	2.4	22.3	1.0	0.4	2.4
Slough	27.7	3.8	12.5	9.1	0.1	2.3
Harrow	26.2	3.7	16.1	1.2	0.3	5.0

Source: 1991 Census Monitors, Table J: Ethnic group of residents.

Clearly, even the small range of examples presented in this section is sufficient to show the high degree of local variation which exists around Britain in terms of population profiles at a single point in time and in the extent and nature of change over time. Moreover, this degree of diversity increases as the grain of analysis is shifted down the spatial scale from regions to counties, districts, settlements and individual neighbourhoods. It is therefore little wonder that there is widespread interest both in having detailed local-area data on population numbers and characteristics and that users are keen that this information is as up-to-date as possible and can be projected into the future with reasonable accuracy.

Selected examples of implications and responses

As well as drawing attention to the diversity of local populations and variations in their pattern of development over time, a central purpose of this book is to explore the implications which follow from them and the responses which relevant agencies make to cope with the issues. Each of the other chapters explores an aspect of local diversity, beginning with the wider demographic considerations of population forecasting and migration and their implications for local government finance and going on to examine specific areas of policy concern. This section highlights the main conclusions of these studies.

In Chapter 2, Dugmore leads off by developing the point made earlier about the strong growth of interest in the local dimension, pointing to the rise of market analysis and geo-demographic targeting over the past decade and the resulting increase in demand for small-area intelligence on population numbers and characteristics. Given the potential for rapid change, these requirements pose a major challenge for developing sound techniques for updating local profiles beyond the most recent census as well as for projecting them into the future. At the same time, however, both the producers and the users of such data must exercise judgement in their work, for instance in their selection of an appropriate projection approach and in their interpretation of the results − judgement which should partly rely on an understanding of local demographic processes and an appreciation of their variability between places and over time.

As demonstrated above, it is the migration component which is the most important determinant of whether population grows or declines at regional and local scale. This is unfortunate for those needing accurate projections because, as Findlay and Rogerson show in Chapter 3, migration has an even greater track record of confounding the experts' predictions than has the birth rate – though Peter Morrison was perhaps going too far in his pithy statement, 'Those who know how to forecast migration know better than to try'. Migration patterns have been becoming more complex over time, responding to a more varied set of stimuli than in the past. In particular, more people appear to have the freedom of moving to where they see the possibility of a better 'quality of life'; and not just the retired and others outside the labour force but also employers and many of their key staff. Given that people differ in their lifestyle and attitudes and in their ability to 'vote with their feet', an in-depth knowledge is needed both of the migration process and of trends in the environment which influence people's migration behaviour.

Local government is one of the biggest users of local population data, and for several very good reasons. One is fundamentally the same as that which inspired the Domesday Book and initiated the Population Census in 1801, namely the need to know what resources can be drawn upon. As Bennett and Krebs show in Chapter 4, the size and composition of a district's population affect both its local tax base and its share of central government grant allocations, with demographic characteristics determining around three-quarters of total local government expenditure. Hence the importance of accurate information not only on overall population size but also on good methods of measuring and monitoring the population's needs. Clear geographical contrasts in levels of local resources, needs and budgetary balance, and in changes in these over time, can be identified, most notably between central city districts, other urban and suburban places, and rural areas and also between London and the rest of the country. At the same time, however, local government spending decisions are influenced by other factors including rising expectations and local political choices.

These latter points are dealt with in more detail in subsequent chapters. In particular, school education forms the single largest local authority service in expenditure terms. As Bradford shows in Chapter 5, it is also one of the areas which has been most affected by recent changes in approach to service provision, with the arrangements introduced by the 1988 Education Reform Act. This change comes hard on the heels of the adjustments required by the massive demographically induced upheavals of the past two decades, which have followed the passage through schools of the 1960s baby boom and subsequent bust. The national fall from peak to trough amounted to 30 per cent over a 12-year period, and in some localities the scale of reduction was over 50 per cent. Bradford shows how local responses to this challenge have varied and goes on to examine the effects of the 1988 Act. Big differences are found in the speed with which education authorities have taken advantage of the Act, which itself – because of the element of parental choice involved as well as the opting-out system – makes for even greater problems of estimating the future level of places needed in state schools.

In terms of the impact of the baby boom–bust cycle, the next major age-specific sector affected after education has been the labour market. This 'demographic time-bomb' began in the mid 1980s, as entry into the fastest period of decline in numbers of 16–18 year olds coincided with the most rapid phase of recovery from the 1979–82 recession. Subsequent return to recession in the early 1990s provided some respite, but as Green and Owen show in Chapter 6, the trend towards lower numbers of school-leavers and the resulting process of labour-force ageing will continue through to the end of the century and beyond. The effect on individual localities will vary enormously, with a few managing to maintain current levels of 16–24 year olds as a result of their very youthful populations or continued in-migration but with the majority experiencing cutbacks of at least 1 in 5 of this age group in 13 years and some of over 40 per cent. Though depending on rates of change in local labour demand, it is safe to say that employers in some places will need to resort more energetically than those elsewhere to finding substitute sources of labour, with the long-term unemployed, older workers, women and the ethnic minorities being featured as the most likely options by Green and Owen.

After labour supply, the housing market forms the next major sector to be affected by the boom–bust cycle. In terms of the pure age structure effect, the peak period of household formation was passed in the late 1980s, so on this basis the 1990s represent a period of decelerating household-formation rates. The latter, when taken alongside the expected high rate of household dissolution arising from the increasing number of very elderly households, would seem to raise the prospect of a significant decline in the need for new housebuilding. Not so, however, says Dave King in Chapter 7. Not only does this 'demographic bulldozer' scenario take no account of the extent of current housing shortages or of the need for replacement of old and dilapidated stock in the future, but it ignores some crucial demographic aspects of the situation. The major cause of extra housing need over the last two or three decades has been the steady fall in household size, resulting from more elderly living alone, more divorces and separations, and more youngsters leaving home as well as smaller family size; and there is no clear evidence of any change in this trend. There is also the factor of mismatch, both in relation to house types and concerning geographical location. In relation to the latter, King points to the strong natural growth in household numbers expected in the shire counties of southern England – a prospect which is already greatly exercising the minds of planners and the NIMBY (Not In My Back Yard) movement.

As noted earlier, the growth in the very elderly population is already with us, as a result of the previous sustained baby boom of the early years of this century. Its implications have been widely researched over the past decade, particularly in relation to the effect on health care spending, social-services support and the growth of specialised accommodation. Corden and Wright, in Chapter 8, focus on one aspect which has not received so much attention: the degree of choice which elderly people can exercise when they go into residential care. This is a rather surprising omission given the importance attached to quality of life nowadays (see Chapter 3) and the 1980s political rhetoric on choice (see Chapter 5). Although showing that people in homes are more mobile than might be expected, this study suggests that elderly people needing residential care have rather restricted choice and that the shift of emphasis from public-sector to private provision over the past few years has not greatly altered this situation.

Several of the themes mentioned in previous chapters recur in Mohan's review of the use of Census data in health care planning (Chapter 9). The National Health Service and the individual health authorities form the second-largest group of users of population projections responding to Joshi and Diamond's (1990) survey. The applications are very similar to those in local government, principally the assessment of local needs for the allocation of central funds to individual authorities and the information needed by each authority in the planning, delivery and evaluation of their services. Despite this, Mohan argues that the potential of the Census is as yet by no means fully exploited. This is partly because it is only in the last ten years that the emphasis in health care planning has shifted down to the identification of localised variations in need. As a result, there remains a substantial challenge to be faced in developing reliable measures of need. The 1991 Census offers particular help here through the results of the new question on limiting long-term illness, but Census data will have to be combined with other sources to build up a comprehensive picture of localised need.

Lastly, if there is one area that truly merits the epithet 'under-researched', it is the contribution of ethnic minorities to local demography. As Robinson points out in Chapter 10, the latest (fairly) reliable picture of the local distribution of the various ethnic minority populations was that provided by the 1971 Census, before the increasing proportion of UK-born members rendered the use of birthplace data obsolete. The 1991 Census question on ethnic group provides the basis for a new assessment, but in the meantime Robinson uses other sources to chart the demographic dynamism of the ethnic groups and examine their distinctive geographical imprints. In so doing, he makes a very clear case for the development of research in this area, recognising the political significance of these groups and the variety of policy issues which are raised.

Further work

Each of the studies in this book, in its own way, reinforces the basic messages that population matters and that a national perspective is inadequate for a range of applications. Admittedly, the examples have been chosen to reflect areas of policy and decision-making where local demographic considerations are paramount, but these areas are some of the most important in the life of the nation today: important in relation to people's opportunities and quality of life, to the national economy and to the long-term sustainability of individual places and their communities, and also important in relation to the amounts of money involved in terms of both current expenditure and capital investment. At the same time, population factors are never going to be the sole determinant of outcomes, so one of the biggest challenges is the accommodation of these with other considerations.

A second clear message from these studies is that the growth of interest in local populations is not going to prove ephemeral. The various topics covered – the growth of the ethnic minority populations and the numbers of very elderly, fluctuations in the numbers requiring school places and entering the labour market, the large scale of spatial redistribution compared with the overall national growth rate, the drive towards more efficient targeting of client groups, the allocation of central funds to local agencies responsible for policy delivery – are of long-standing nature and seem destined to remain important for the foreseeable future. It is for this reason that this book is seen as a beginning rather than an end in itself.

Timing is important in this context, because the next five years are going to provide hitherto unparalleled opportunities for the study of local populations. This is due not just to the steadily growing interest in this topic but also to the massive increase in data availability expected over this period, primarily due to the release of the various data sets deriving from the 1991 Census. These comprise not only the fullest set of local-area tables ever produced from the Census, but also the special tables on migration and journey to work, the incorporation of 1991 data into the OPCS Longitudinal Study, and the first-ever Sample of Anonymised Records. Other sources can be used to provide in-depth understanding of individual circumstances and behaviour, though not at such a fine spatial scale; these include the latest wave of the National Child Development Survey and the recently started British Household Panel Study, as well as continuing sources such as the registration of births and deaths, the NHS Central Register on internal migration, the General Household Survey and the Labour Force Survey.

To prompt readers into their own investigations, the following chapters highlight some of the ways in which these data sources can be used to update the results reported there and to follow up research questions posed. For instance, Mohan and Robinson comment on the value of the information produced by the new questions on limiting long-term illness and ethnic group respectively. Corden and Wright point to the need to update our knowledge of the proportion of elderly people living in residential care and examine their experiences of moving into and between homes, particularly in the context of the new arrangements for community care. The studies by Bradford, Green and Owen, and King each offer opportunities for examining the detailed local patterns of changing age-group size between 1981 and 1991, for exploring the extent to which this has influenced policy and planning and for using the results of analysis to refine future projections. Dugmore cites the value of the detailed 1991 Census results for checking the accuracy of current estimation and projection techniques, as well as recognising the importance of this information for the more refined targeting and allocation procedures needed both for more effective marketing and in the public-sector areas covered by Mohan and Bennett and Krebs. In terms of projections, the key to better projections lies particularly in the migration area examined by Findlay and Rogerson; already a substantial research agenda has been identified for this (see Champion, 1992). As outlined in the preface, the purpose of this book is to provide a springboard for this research activity.

2. POPULATION PYRAMIDS AND SHIFTING SANDS:

Targeting future investments to give the changing British public their just deserts

Keith Dugmore

Many organisations, both public and private, now use information about local populations to target their services to the public. As mentioned in Chapter 1, market analysis and geo-demographic targeting has grown rapidly in the last decade. This is not surprising given the extent to which population characteristics differ between places, not least in age structure (often represented by demographers as a pyramid) as well as in household type and social composition.

Comparatively little attention, however, tends to be paid to the extent of the changes taking place over time despite the fact that the pace of change at the local scale can often be considerable, even over relatively short periods. It can sometimes prove dangerous — and inefficient in marketing and targeting terms — to assume that the population structure of an area is more or less the same as at the previous Census, perhaps 5–10 years ago. For making investments which may have a payback period of 20–40 years, it is, of course, even more vital to anticipate trends.

This chapter builds on the observations of major regional and local variations in population characteristics and trends made in Chapter 1. It begins by identifying the types of organisations for which knowledge and understanding of these variations is crucial for their current activities and planning decisions. It goes on to examine the sort of information which is needed about future population developments and to outline the alternative projection methods available and how to select the most suitable approach. The chapter then presents a case study in small-area population projection, assuming a relatively common set of information needs, and ends by stressing the importance of the arrival of 1991 Census data for this work.

At the outset, some definitions should be clarified. The term 'projection' is used in this chapter to mean a calculation based on stated assumptions. It is used in preference to both 'forecast' and 'prediction'. Benjamin (1968) has argued that forecast implies the expectation of fulfilment; 'prediction' implies this even more strongly. Extrapolations, which assume no change in current trends, are a sub-set of projections. The term 'small area' is used to refer to areas which are smaller than counties, and usually smaller than a typical local authority district of 100,000 people. Population change in such areas can be very rapid.

Who needs small-area population projections?

All organisations which provide services to populations are likely to need projections for small areas. An increasing number of organisations − but by no means all − are now using current information in their planning. Few, however, appear to be looking even as far as five years ahead, although the recent fascinating review by Joshi and Diamond (1990) suggests increasing interest, particularly among local government and health care organisations.

It is still common for strong distinctions to be made between public sector and private sector organisations, but both provide services to populations. One may be concerned with 'public services' and another 'consumer markets', but both share the need to make decisions about resource allocation and investment in facilities. Whether reviewing the allocation of staff to a sales territory or to a social work patch, or investment in a bank branch or a health clinic, all decisions need to be underpinned by an understanding of current demographic structure and likely future trends.

The size and composition of population and consumer markets can change markedly within a few years. Even when looking at the total population for areas as large as counties, there were significant changes between the 1981 and 1991 Censuses (Table 2.1). In a county such as Surrey, which experienced virtually no change in total population, the age composition nonetheless altered markedly (Table 2.2).

Turning to geographical subdivisions, change in total population is rarely equally distributed across a county. Dividing Humberside into its constituent local authority districts, for example, we see changes ranging from −8.7 per cent to +11.4 per cent (Table 2.3). For smaller areas, such as service catchments, wards or postal sectors, variations will often be even larger. Thus, the size of a particular population segment within a small area can − indeed, is likely to − alter considerably within a single decade.

In practice, resource allocation and investment decisions are taken by many different types of organisation. Needs for demographic information vary, as the following examples illustrate.

Table 2.1 Population change 1981−91, selected counties

County	1981	1991	% change
Cambridgeshire	567,891	628,179	10.6
West Sussex	650,124	683,318	5.1
Surrey	991,074	989,370	−0.2
Tyne & Wear	1,135,492	1,073,628	−5.4
Merseyside	1,503,120	1,366,884	−9.1

Note: resident population on 1981 base to allow direct comparison.
Source: 1991 Census County Monitors.

Table 2.2 Population change by age, 1981–91, Surrey

Age	% change
0–15	−9.7
16–29	0.3
30–44	8.1
45–PA	−2.5
PA–74	−7.7
75+	26.9
All ages	−0.2

Note/source: see Table 2.1. PA = pensionable age (65 for men, 60 for women).

Table 2.3 Population change 1981–91, districts in Humberside

District	1981	1991	% change
Boothferry	59,880	62,868	5.0
Cleethorpes	68,325	67,826	−0.7
East Yorkshire	73,980	82,433	11.4
EYB of Beverley	104,925	109,476	4.3
Glanford	65,959	70,634	7.1
Great Grimsby	91,541	88,758	−3.0
Holderness	45,863	50,121	9.3
Kingston upon Hull	266,760	247,978	−7.0
Scunthorpe	66,047	60,277	−8.7
Humberside total	843,280	840,371	−0.3

Note/source: see Table 2.1. EYB = East Yorkshire Borough.

School rolls

This requirement is typically for a relatively small catchment of one or two miles' radius, and for projections of the child population by single years of age. It is usually assumed that a high proportion of the child population will use the school.

Retail superstore

In this case the catchment is often defined as a drivetime isochrone of perhaps 15 minutes. The catchment is unlikely to have any respect for administrative boundaries. The retailer will be concerned with the numbers of people in particular segments of the population (for example affluent suburbanites) and will only expect to attract a proportion of the total potential market to the store.

Police subdivisions

Police forces usually divide their county areas into a number of sub-divisions, twenty being typical. Sub-division boundaries might not be changed for several years, but the numbers in the major client group – young males – can alter considerably. There is often a case for reviewing and redirecting police resources from one sub-division to another, or, more radically, revising boundaries as populations change.

Health
Both public and private health care organisations need information for administrative areas for service catchments. Target populations such as infants, children, young women and pensioners can be specified, and there can also be interest in such classifications as marital status, household composition and ethnic group.

Leaflet distribution
The use of targeted marketing has increased rapidly in recent years. Many campaigns use local door to door delivery of leaflets. This requires good information on both the social profile and household numbers in small areas such as postal sectors.

In many cases the power and benefits of geo-demographic targeting can be increased by having information not only on the total population and its age distribution, but also such characteristics as housing tenure, social class and car ownership.

What information is required?

The specification of what projections are to be produced must be determined at the outset. There is a distinct danger of being seduced by interesting methods and losing sight of the target output or — to use that ungainly but accurate word — 'deliverables'. The user and producer need to agree answers to a series of questions.

How much subject detail?

Projected population might be classified not only by sex and age (broad or detailed bands) but also a variety of other variables such as economic activity, ethnic origin and social class. Whilst in principle some users might ask for considerable subject detail, most are likely to settle for broad categories which they might well view as better than no projections at all. One merit of producing more detailed classifications is that they can then be used as flexible building bricks to produce non-standard categories such as 11−16 year olds. This pursuit of detail is, however, in practice often ameliorated by the fact that only a limited proportion of a particular population actually uses a service — the incidence of take-up rates can be very important

How much geographical detail?

The increasing use of demographic information through the 1980s points to more and more interest in targeting small geographical areas. The key issue here is that the areas can be specified in an infinite number of ways, few of which correspond to standard administrative definitions. Neither leisure centre catchments nor drivetimes to a DIY store conform to administrative boundaries.

The obvious strategy to give geographical flexibility is to produce projections for small building blocks which can then be aggregated to produce larger areas on an *ad hoc* basis.

How far ahead?

This touches on the heart of the reason that the user has shown interest in projections in the first place. The user wants to know what an area will look like in 2, 5, 10 or 20 years' time. The demographer will need to temper this by pointing out that for small-area projections experience suggests that looking five years ahead is usually reasonable, but to exceed ten years is probably unrealistic. On the question of time horizon, it is also worth bearing in mind that many users are not locked into years particularly favoured by demographers: they are just as likely to ask for 1994 and 1997 as for 1995 or 1996. It is usually advisable to produce projections for every year in the time period.

One projection or a range?

Thompson (1974) has argued in favour of using more than one set of assumptions and so producing a range of projections rather than a single one. A single projection can be misleading for, however many caveats any commentary makes, it is probable that users will accept the projection uncritically. A range gives the user an idea of the unreliability of the projections.

Gilje (1974) has argued in favour of a range of four alternatives. If only two are provided, the user will probably split the difference; if three are provided, the middle one will almost certainly be chosen. Another practical reason for producing four alternatives when using the demographic components model is that fertility and migration are much more variable than mortality; alternative 'high' and 'low' assumptions (chosen in the light of past experience) for the two variable components will produce a range of four projections.

The use of alternative assumptions relieves the demographer of some of his burden at the expense of the decision-maker, who will have to choose which projection to use. The use of alternative assumptions, however, can easily get out of hand: high and low fertility, mortality and headship rate assumptioms will produce a range of eight alternative projections.

Which projection method is best?

This is an important issue because a variety of alternative methods can be used to produce population projections. The choice of approach should be influenced by several factors. Obviously very important is the type of information required, as outlined above, including aspects such as subject detail, geographical detail, time horizon and whether a single projection is needed as opposed to a range of alternatives. The choice of method will also be determined by very practical considerations of data availability, budget and deadlines. There may also be significant local conditions to take into account, for example the effects of major redevelopments such as those taking place in the London Docklands in the 1980s.

As regards the variety of projection methods, there are at least three demographic approaches, distinguished basically by level of detail and sophistication. The simple ratio method assumes that future changes in a small area will be proportionate to those in a larger area. Secondly, there can be simple extrapolation of observed trends, for example total population, socio-economic group or ethnic origin. More detailed projections are based on an analysis of past data. This is not necessarily the simple extrapolation of the most recent trends. It can be applied separately to the components of demographic change and can, for example, involve alternative high or low assumptions about fertility and migration, based on historic data.

Two other widely used methods are housing capacity models, which assume that the population is determined by the stock of housing that will be available, and employment base models, which relate changes in population to changes in numbers of jobs. Further information on such alternatives can be found in Woodhead (1985).

In terms of data sources, four broad categories of data can be identified. The successive decennial Censuses give aggregate statistics for small areas. It is possible to classify the population at a series of cross-sections in time. The Census also includes a question on 'address one year ago', which enables analysis of migration in the pre-Censal year. Secondly, aggregate official statistics of births and deaths by ward enable the analysis of fertility and mortality. Administrative lists can (if they are available!) be used to provide information on population total and, in some cases, migration. Under this heading the vaccination file (for under-5s), school rolls, the electoral roll, the National Health Service Central Register, and water and electricity company customer files all offer some scope. Lastly, plans for future investments in housing and jobs can often be of considerable local significance.

In the final analysis, individuals will have their own personal views on the best approach for a particular task. My own checklist of steps and considerations is as follows:

- start with simple models and get some early results
- adopt a critical attitude, challenging the plausibility of the results
- don't be afraid to mix and match models, adopting a hybrid approach
- think long and hard before introducing complex models with onerous data requirements: massive data inputs do not necessarily produce much better outputs
- remember that gross migration flows can vary considerably from year to year: the Census 'change of address' year of 1990/91, for example, was a period of exceptionally low overall mobility
- remember that whilst the use of housing stock models might be convincing in rapidly growing Milton Keynes, many inner urban areas have experienced huge population declines in a static housing stock.

A case study in small-area population projections

Organisations' needs

The following case study illustrates the issues and approaches outlined above, drawing on experience in helping to develop small-area population projections at CACI Limited. This experience is relevant to many organisations, ranging across retail, finance, property, direct marketing and public services. All these are interested in using small-area information to target resources. Some have limited requirements of, for example, Census counts. Others, however, wish to draw on a wide range of data sources, and place a premium on information being as up to date as possible.

This exercise was undertaken in response to the growing number of organisations expressing interest not just in current estimates but in how trends might extend into the future. For many organisations, the wider use of demographic information during the 1980s had brought massive benefits. Many more decisions now have a quantified scientific input rather than being based on past custom or gut feel. Banks, for example, are increasingly sophisticated in researching branch performance. Retailers seek out the most profitable markets. Bearing in mind the scale of these operations, the benefits to be obtained from an informed guess about future population changes will usually justify the cost.

It was recognised that, for most organisations, the basic interest is in looking at population classified by age and sex in five or ten years' time. The projections produced by central government are very useful in some instances, but their major drawback has been that information is available only for counties or, in some cases, local authority districts. Some organisations, however, need information for smaller areas, whilst many are concerned with *ad hoc* catchments or different geographies such as postal districts.

Previous experience suggested that most users have realistic expectations of what projected information can be produced. In some instances a requirement for detailed information about a small geographical area might justify a special project, as can an exceptional area such as Docklands. Generally, however, the need is for a broad strategic view of the future which is plausible on the basis of recent trends, though this does not discount the need for judgement.

The information produced

Taking into account the range of clients' requirements of small-area population data, it was decided to produce a standard set of projections to the following specification:

- *Population.* Usually resident (Census definition) by sex and quinary age bands to 80+.
- *Geographical coverage.* It was decided that the small-area projections should cover the whole of Great Britain.
- *Geographical detail.* Small building blocks were essential to give geographical flexibility in terms of output areas.
- *Time horizon.* Approximately ten years ahead, with interim years also being made available.
- *Alternatives.* It was decided to produce a single projection as standard, with the exploration of alternatives being treated as special projects.

Method used

Several methods of producing projections were considered. It was concluded that a hybrid method incorporating trend extrapolation, demographic and ratio models would be most appropriate.

The essence of the chosen method is to produce updated estimates of the population by sex and age for small areas for each year following the 1981 Census up to the 1987 base year. These are consistent with government mid-year estimates. Having updated the estimates, these 1981–87 trends are projected forward, constraining the results to those produced for local authority areas by central government. These two basic steps are described below.

Updated estimates

At a ward level the model incorporates data on actual births and deaths, recorded by Office of Population Censuses and Surveys (OPCS) and General Register Office (Scotland) [GRO(S)], and changes in the electorate, derived from summaries of the registers compiled annually by each local authority (Bracewell, 1987).

The base population has been aged using a single year cohort survival model. This approach generates a time series with consistent results for each year since 1981. The vital statistics are used to add births and subtract deaths from specific age cohorts.

Electoral data is used to model the effects of in- and out-migration. For example, the completion of a new housing development can cause a net inflow, whilst an inner city clearance scheme can result in a net significant out-flow.

The age structure of migrants is calculated by applying age–sex specific propensities derived from Census data for that locality. A series of cross-checks is made with individual authorities to determine whether the changes in the size of electorate are genuine. Adjustments are required to account for boundary changes or changes in electoral practice.

Within each district the unconstrained ward estimates are controlled to the official mid-year estimates using an iterative procedure which takes account of the variations in age structure across wards. The operation of this controlling algorithm is monitored and in practice the differences measured at this stage are marginal. A differences table is used to move between the definition of the home population (adopted by the Registrar General when producing mid-year estimates) and the resident population (as used in the Census), making the necessary adjustments for students and the armed forces.

A further model is then used to split the ward populations down to individual Enumeration Districts (EDs) for re-aggregation to *ad hoc* areas and other non-Census geographies. The ED model is similar to that used at a ward level, but does not take into account specific migration flows. Estimates of births and deaths (produced from the ward statistics) are used to age the base population. The migration component is then defined as the difference between the constrained ward population and the sum of the EDs. The model also ensures that the resulting age profile is reasonable in terms of the ratio of adults to children. The population resident in those institutions defined as special EDs is held constant. A second controlling algorithm is used to ensure the ED populations within a ward are consistent with the ward total by age and sex.

Projected trends

Using the updated population base thus derived (relating, in this example, to 1987), the projections themselves are produced in four stages. First, ward population totals are projected, using the trends established in the updated estimates. Here, more recent years are given a greater weighting than earlier years; areas of rapid growth or decline are assumed to change at a decelerating rate. Next, age profiles are obtained by initially classifying wards according to total population change and population density; trends for different classes of ward are then projected forward. The resulting ward projections by age and sex are constrained to those produced by central government and to the most recent mid-year estimates; this process again involves using the differences table to move between definitions of home population and resident population. Finally, a further model is used to spread the ward estimates down to ED level, reflecting the ratios at the time of the 1981 Census: this enables re-aggregation to *ad hoc* areas and non-Census geographies.

Some illustrations of results are given in the form of a map of overall population change (Figure 2.1) and a version of a population pyramid that shows percentage change in the size of age groups (Figure 2.2). Figure 2.1 shows projected population change between 1990 and 2000 in the postal sectors of East Anglia. Such information could be particularly valuable to a

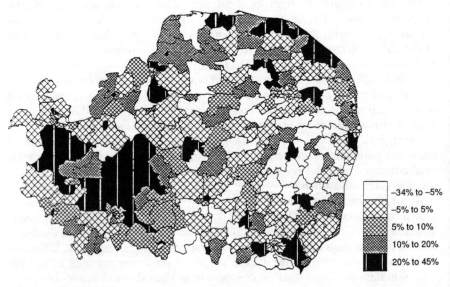

-34% to -5%
-5% to 5%
5% to 10%
10% to 20%
20% to 45%

Fig. 2.1 Projected population change 1990–2000, postal sectors in East Anglia (*Source*: Woodhead and Dugmore, 1990).

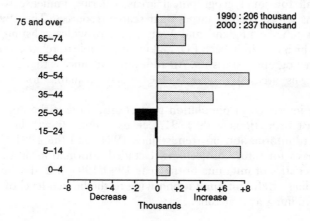

a) Bracknell: 15 minutes drivetime from centre

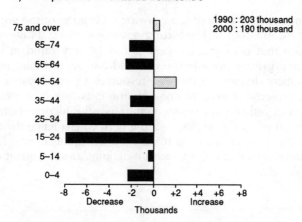

b) Liverpool: 2.5 mile radius from centre

Fig. 2.2 Projected population change 1990–2000 for two *ad hoc* areas: A: Bracknell: 15 minutes drive from centre; B: Liverpool: 2½ mile radius from centre (*Source*: Woodhead and Dugmore, 1990).

retail group considering where to open superstores in future years. What might at present be a finely balanced decision between two alternative sites in different parts of the region might become far simpler on looking at prospective population growth or decline. Figure 2.2 contrasts projected change in numbers by age for a 15-minute drivetime area around Bracknell with that for central Liverpool: a fashion chain, for example, might well use this information when deciding which of its existing stores to refurbish.

The arrival of the 1991 Census

The publication of statistics from the 1991 Census brings many opportunities for demographers. The Census provides up-to-date information in great subject detail for small geographical areas, offering immense scope to all organisations which need to target their scarce resources. The 1991 Census, for the first time in England and Wales, also provides a listing of all the postcodes which lie within each ED, making it possible to relate administrative records to demographic statistics. An additional advance is that digital boundaries of census areas are being made widely available for analysis and mapping.

For those interested in population projections, there is the opportunity to compare past projections with 1991 Census counts. It will be possible to compile more information on trends since 1981, and the 1991 Census will provide a basis for future projections. Detailed information on the numbers and characteristics of migrants in the year 1990/1991 will also be available, though needing careful interpretation owing to the limited level of residential mobility pertaining at that time.

Conclusions

It is heartening to see that there is a growing perception of the importance of demographic information in supporting decision-making. However, there is less recognition that demographic change can be very rapid in local areas. Both public and private organisations should anticipate future trends when making investment decisions, otherwise resources are re-allocated by default. Producers of projections need to consider the importance of meeting users' needs (a demand rather than supply emphasis), the merits of starting with simple models to get early results, and the need to challenge the plausibility of results. Lastly, projections need to be used with judgement: they are just one part – albeit a vital one – of decision-making, as subsequent chapters in this book make clear.

3. MIGRATION, PLACES AND QUALITY OF LIFE:

Voting with their feet?

Allan Findlay and Robert Rogerson

Migration plays a major role in producing changes in the size and composition of local populations. As shown in Chapter 1, the net level of migration inflows and outflows is more important than natural change as a component of overall population change, and its importance tends to be more important at the local scale than at broader regional and national levels; partly because most movement takes place over shorter distance. Even where there is an approximate balance between inflows and outflows, significant changes can take place in the local population profile because the selective nature of migration may mean that the newcomers to an area may be very different from those leaving it.

The central message of this chapter is that migration patterns are becoming more complex and that migrants are responding to a much more varied set of stimuli than in the past, leading to greater selectivity in the types of places people choose to move to. In particular, people are giving more attention to 'quality of life' considerations at the expense of economic, or indeed strictly employment, factors — something that is nearly as true for longer-distance movements as for the shorter-distance residential mobility that has traditionally taken place for housing and environmental reasons. The chief implication for local planning is that, in order to keep up to date with changing local population profiles and to anticipate trends over future investment periods, detailed monitoring of migration is needed, as also is an accurate understanding of the processes which influence the movement of people between places.

People and place

The relationship between people and place has been a longstanding interest of geographers. Population geographers have laid particular emphasis on interpreting the effects of fertility, mortality and migration in producing patterns of population change. In modern Western society, the decline of both fertility and mortality has in recent decades resulted in greater attention being given to migration as the primary process re-distributing population between one place and another.

The shift in the importance of migration as a critical determinant of the population structures of places has occurred at the same time as people's attachment to place has been transformed by the infusion of new values in society. There has been a renewed emphasis on the locality as a focus for spatial attachment and support. Migration has come to be recognised as one of the key societal processes linking people and place (Champion, 1992). The manifestation of these trends has meant that commentators such as Gordon (1992) have found it increasingly difficult to apply traditional macro-economic models of migration due to their omission of powerful cultural, social and political forces. At the same time, innovative investigations by Fielding (1992a, b) have struck a resonant note with other social scientists in highlighting the intimate and reciprocal relationship between migration and social change. To quote Fielding (1992b, p. 247) 'the connections between class and regions, society and space, are increasingly visible, and as this happens, the centrality of migration in these relationships and processes is becoming fully revealed'.

For some, the changing role of capital and the restructuring of society are viewed as determining agents in understanding migration. For others, culture, capital and agency intersect to produce a somewhat more complex, flexible and less constrained migration system. In investigating the relationship between migration and quality of life, this chapter assumes the latter position, arguing that while the positioning of a potential migrant within society affects the propensity to migrate, the migration act itself has become more flexible and less spatially constrained.

Moreover, the increased potential for migrants to choose where to live has been mirrored by an increased potency amongst those who control information about places to project 'new' images of places and to reposition localities within the migration system. While some academic research (Ashworth and Voogd, 1990) suggests that 'place marketing' has not been entirely successful, the evidence from Britain in the 1980s would suggest that different population groups have been targeted by local authorities. For example, the 'Glasgow's Miles Better' and the 'Bradford Bounces Back' campaigns portray a different image from those campaigns publicising postmodern architecture such as in the London Docklands (Zukin, 1992). The selective nature of these projects cannot have been other than to have a differential impact on migration.

This increasing local complexity in migration patterns has been ably demonstrated by analysts of migration in Britain in the past decade. In the early 1980s, great attention was given to the consistent regional patterns of net losses from Scotland, the North, North West, Yorkshire and Humberside and the West Midlands of England and the consistent gains recorded by the East Midlands, South West and East Anglia regions in England and in Wales. Only in the South East of England did the position change during the first half of the decade, from net in-migration to net out-migration. By the late 1980s and the early 1990s, in contrast, the patterns had become much less clear, with place-specific characteristics apparently becoming more dominant in influencing migration flows and with a mosaic of migration gains and losses displacing the clear cut regional divides of the past (Rees, Stillwell and

Boden, 1992). Consequently, the use of standard economic regions to analyse migration fails to show the complexity which exists at other geographical scales.

Against a backdrop of only very low population growth at the national scale, the pattern of population structures at the local scale across the country has shifted dramatically, to produce what Champion and Townsend (1990, p. 67) described as 'a mosaic of both absolute growth and absolute decline across Britain with relatively few places maintaining a static population size'. It is in relation to this shifting pattern of population that migration emerges as one important aspect of the relationship between people and places. This is important in terms of what migration patterns tell us both about the changing significance and positioning of places within Britain and about the factors influencing the migration process, particularly the motivations of the actors involved.

Given this changing relationship between people and places, one way to analyse the shifts which have occurred is to re-examine migration motivations. The remainder of this chapter draws on the results of a recent attitudinal survey of migrants which sought to consider one aspect, namely the role of quality of life in migration motivation and information choice. Using the responses to the survey, the chapter considers the degree to which quality of life issues should be viewed as important in understanding and explaining recent migration trends, and concludes by outlining the implications of these findings for those involved in charting trends on the size and composition of the population of individual places.

A survey of migrants' motivations

In order to interpret the complex local patterns of population change which have characterised Britain in recent years, a national attitudinal survey of migrants was carried out with the aim of revealing the origins and destinations of those involved in population relocation and of identifying migrant perceptions of the factors which underlie their moves. The survey was commissioned by the authors based in the Glasgow Quality of Life Group and was carried out in December 1989 by a London-based social survey company. Like all nationwide surveys it is subject to sampling error, so interpretation requires caution.

A sample of 2,225 adults was selected from Britain's adult population (16 years or over) stratified by age, sex and class and partitioned between the major standard economic regions in proportion to their resident population. A migrant was defined as a person who, at the time of the survey, had moved since 1984 from a different city or region. Intra-urban and short-distance rural migration was therefore excluded from the analysis and the focus was on inter-city and inter-regional mobility. Of the 241 respondents who had migrated during the defined five year period, 29 per cent had moved between cities within one of the ten standard regions, while the remainder were inter-regional migrants.

As Table 3.1 shows, in response to questioning about their motivation for selecting their chosen destination, most of these migrants indicated that, at

Table 3.1 Important aspects in migrants' choice of living environment

Reason	% respondents indicating reason was important or very important in migration decision		
	Inter-regional migrants	Intra-regional migrants	All migrants
Employment opportunities	58.6	48.1	55.2
Living costs	45.1	49.4	46.5
Quality of life	68.5	75.9	71.4
Family/coupling reasons	40.2	59.5	46.5
Number in sample	162	79	241

Note: migrants are defined as people moving between cities.
Source: Authors' survey (see text).

least in part, their decision as to where to live was conditioned by aspects of quality of life. Over 70 per cent of all migrants said that quality of life had been 'important' or 'very important' to them in their decision-making process. Employment-related issues were of less importance (55.2 per cent), as were other economic factors such as living costs. These results clearly do not conform to the traditional interpretation of inter-regional migration as being dominated by labour market criteria.

Given the surprising nature of these results, it is important to emphasise that, although the number of migrants is small, the sample does appear to be a representative one (Findlay and Rogerson, 1991). This can be demonstrated from the sample's demographic and social profile which exhibits the characteristics to be expected of longer-distance movers. Thus, 52 per cent of the migrant population sampled were between 18 and 34 years of age while 12.4 per cent were over 65. The overall age profile is consistent with the age-specific migration probabilities which one would expect from other migration surveys, with a peak level of mobility amongst recent entrants to the labour market and those in the early stages of family formation, and a migration trough for people in the older part of the workforce. Mobility levels rise again in retirement as one would expect in contemporary Western society (Stillwell and Boden, 1989; Warnes and Law, 1984; Warnes, 1990). Concerning social class, the sample was also in line with the results of other research. Professional and managerial groups formed a larger segment of the migrant sample than in the population as a whole. Migrants were less likely to be drawn from skilled, semi-skilled and unskilled manual workers.

Quality of life as a determinant of migration

Why should it be that quality of life considerations are of growing importance in influencing migration and local population redistribution between places?

One reason concerns the changing age structure of the population. During the 1980s, the size of the younger cohorts of the population entering the labour force for the first time became smaller relative to the number of older persons being shed from the labour force either through retirement or redundancy. Similarly the ratio of people of pensionable age in the British population reached the highest level on record by 1988 (18.2 per cent of the total) and within the next 30 years the proportion will rise to over 21.0 per cent. The result of these demographic trends, setting on one side for the moment the effects of other social and economic trends, would have been to reduce the volume of job-related moves of new labour market entrants, while at the same time swelling the cohorts of people likely to be involved in retirement migration and non-labour market moves.

The 1980s were also the time when for economic and technological reasons population concentration in the old conurbations was no longer as necessary as it had once been. Keeble (1989) has discussed the economic forces which underlay the shift in industrial production to smaller units, whilst the shift in employment away from manufacturing to service-based sectors has also reduced the need for the central urban locations once demanded by industry. This in turn has permitted the managers of small firms to be more flexible in their locational decision-taking in favour of rural areas. Limited empirical evidence (Keeble and Gould, 1985; Bolton and Chalkley, 1989) supports the interpretation that 'a major compelling force appears to be selective migration by managers, professionals and more highly qualified workers and their families from cities to smaller towns and villages, largely for quality of life and the environmental reasons' (Keeble, 1990, p. 240). Based on a log linear analysis of United States migration, Williams and Jobse (1990) have claimed that quality of life rather than economic factors hold the key to explaining United States migration, and that analysts should see jobs as facilitators of migration rather than as being causal; others, however, dispute Williams and Jobse's conclusion (e.g. Greenwood and Hunt, 1989; Clark and Cosgrove, 1991).

Several researchers (Hepworth, Green and Gillespie, 1987; Goddard, 1991) have investigated the effects of new technologies such as the computer revolution in changing employment structures. By 1984, it was estimated that 47 per cent of employment in the UK was in occupations primarily engaged in the production, processing or distribution of information, and that the geography of this new information economy was responsible for fundamental restructuring of the relationship between core and periphery in the UK economy. Goddard's (1991) statistics reinforce the view that a significant regional divide exists across the United Kingdom, similar to that identified by Stillwell, Boden and Rees, (1990) in relation to migration patterns. Goddard notes, for example, that 52 per cent of employment in the south of Britain was in the information industries, compared with only 42 per cent in the periphery.

Distribution innovations have permitted new ways of delivering goods and services to peripheral markets. Developments such as facsimile machines have limited some of the constraints of physical distance through the profound effects which they have had on the organisation of retailing, banking, insurance and publications (Gillespie, 1987). Information-based production systems have also been released by telecommunication and computing innovations from locating labour and capital together. The networking of computers has made it possible to transport effectively labour power almost instantaneously over long distances. This has given organisations and individual employees much greater locational flexibility. Similarly, telecommunications has permitted the increasing spatial separation of management from production in new regional and international spatial divisions of labour (Massey, 1984; Castells, 1989). This has meant that management has been increasingly able to consider non-economic issues in their locational decision-making and to optimise 'soft infractructure' factors (Blakely, 1987) such as proximity to attractive physical and social environments.

Greater spatial flexibility has also been stimulated by the profound changes taking place in the domestic sphere, notably involving the undermining of the concept of the suburban home as social 'refuge'. Kirby (1990) has noted that the suburban home has come under severe strain due both to the erosion of the family unit (Lelievre, 1991) and to the impact of rising levels of urban crime, stress and congestion. The home has been transformed into an 'office space and communications centre, a long term food processing space, a locus of complex electronic equipment, even a refuge from potential nuclear war' (Kirby, 1990, p. 5). As a result of the assault on the home, Kirby suggests that the individual has increasingly looked to the locality of residence as a key dimension of his or her life. Thus, quality of life in the locality has become an important issue on which residents express strong opinions (Clarke and Kirby, 1990).

The 1980s were also the decade when individualised lifestyles became an increasingly apparent aspect of British society. This has been interpreted as just one further aspect of the evolution of the so-called post-modern condition. For some people this drive led them to abandon what they perceived to be the urban rat race (Bolton and Chalkley, 1990), while for others the search for a good quality of life was more positively associated with the choice of an alternative residential location which was more in tune with their personal values systems (Fielding, 1992a). Equally, for some people, this has meant longer-distance commuting from work to home, in order to allow a choice of a more desirable residential location (Parr, 1987; Fuguitt, 1991). Others have opted to relocate in the city centre where their aspirations for particular types of lifestyle and processes of consumption can be met (Day and Walmsley, 1981). This gentrification trend has been widely observed in many different cities (e.g. Smith and Williams, 1986; Smith, 1989).

There would therefore seem to be a number of forces, ranging from demographic through economic to societal and cultural ones, which favour migration to locations which, in terms of the migrants' own views, might be thought of as offering a better quality of life. To quote from the *Economist's* view of the *World in 1991*: 'People like having cash in their pockets, but many, including the well-off, believe that the quality of life in Britain — the cleanliness of its cities, the quality of its education, the punctuality of its trains — has been deteriorating, is deteriorating and should be improved' (*The Economist*, 1990, p. 27). Furthermore the CBI Relocation Unit has noted that 'quality of life and effects of any proposed relocation on the family life of individuals is rising in importance' (CBI ERC, 1991, p. 20).

Quality of life and migration: a cautionary note

While it is possible to present a range of arguments that helps to account for the growing role of quality of life in explaining migration behaviour, it is much more difficult to determine actually what proportion of current population moves are affected by this 'vague ethereal entity' (Campbell, Converse and Rodgers, 1976). Previous behavioural research on other aspects of migrants' motivations have encountered a myriad of methodological and philosophical problems. For example, it should be self-evident that no one factor operates in isolation from all others in influencing migration behaviour, and also that individual migration decisions like all human choices are made within a highly constrained societal and economic context. These limiting factors should not be interpreted as precluding meaningful analysis of the positive motivating factors which underpin the human decision to migrate, but they necessitate a careful specification of the realms within which decisions are made, lest the role of individual choice and motivating forces are over-stated.

Migrants clearly make decisions about moving in relation to a range of very different types of forces. With regard to any discrete migration move, several different and perhaps unrelated factors may come into play. It would therefore be wrong, for example, to suggest that all migrants leaving Scotland in the 1980s were motivated only by employment factors or that all out-migration from the South East was motivated by quality of life factors. Champion (1989b, p. 58) expresses this in a different way when interpreting aggregate patterns of British migration: 'population redistribution across the settlement hierarchy is not a simple phenomenon with a single direct cause, but rather is a tug of war between alliances of centripetal and centrifugal forces'. Quality of life factors should therefore be interpreted as one of several forces influencing the migration process at both the individual and aggregate levels of analysis.

Valid individual answers to the question 'why did you move?' may disguise the fact that one decision-maker in a household will usually have more influence than others in determining the timing and location of a move. With migration, as with all decisions, the final outcome is also highly dependent on the quality of information available to the decision-maker(s). Thus, even were quality of life to be acknowledged as a dominant influence on a migrant household, there would inevitably be discrepancies between the actual locations which might be shown by socio-economic indicators to offer a desirable lifestyle and those selected by migrants. These differences would be unavoidable because of the imperfections in migrants' knowledge of both available quality of life and migration opportunities.

It is also relevant to note that even were quality of life the most critical influence affecting migration, one would expect a differentiated migration pattern reflecting the varied quality of life definitions of different groups in the population. For example, the definition of quality of life varies with age and position in the life cycle. As Table 3.2 indicates for two population groups – the elderly and 25–34 year olds – some features such as good health provision and low crime rates make an area attractive to most people,

Table 3.2   The ten most important dimensions of quality of life: the views of 65+ and 25–34
year olds

Dimension	%
65 and over	
1. Health service provision	93.3
2. Violent crime levels	91.7
3. Non-violent crime levels	91.2
4. Cost of living	90.8
5. Pollution levels	86.8
6. Shopping facilities	81.6
7. Access to areas of scenic beauty	75.8
8. Climate	68.5
9. Owner-occupied housing costs	58.5
10. Quality of council housing	55.5
25–34 year olds	
1. Education facilities	93.9
2. Health service provision	93.2
3. Violent crime levels	90.7
4. Pollution levels	88.8
5. Employment prospects	88.1
6. Wage levels	88.6
7. Non-violent crime levels	87.9
8. Cost of living	87.2
9. Shopping facilities	82.2
10. Owner-occupied housing costs	81.6

Note: data refer to the percentage indicating that the dimension is important or very important.
Source: Authors' survey.

but others are age-specific in their significance. For example, educational and
employment-related indicators (such as education facilities, wage levels and
and employment prospects) are included in the 'top ten' factors by the
younger age cohort, with less emphasis on cost of living and housing. In
contrast, the elderly attach more importance to cost of living, shopping and
environmental concerns. Given this position, it is possible to resolve the
apparent paradox that quality of life reasons could account simultaneously
for migration flows both into and out of the same region (but for different
groups in the population).

A final qualification which must be noted is the scale at which the reasons
for moving come into play as explanations of the migration act. It is quite
conceivable that many of those who left Greater London to live in other parts
of South East England in the 1980s did experience a real improvement in
their quality of life. This does not mean that quality of life was either the key
stimulus which initially set in motion a relocation move, or that the migrants
involved in the move optimised their choice of destination in terms of quality
of life. It may have been perceived as the key factor in choosing between the
two or three possible destinations offered to an individual by a company
instructing an employee to change branches, or it may, for example, have
been the key issue in selecting which of several possible residential areas was
selected for house purchase within a larger labour market area. In either of
these examples, the individual migrant was acting on quality of life issues at

one level in the decision-making process while choosing to ignore those aspects of the relocation decision which lay beyond their control in terms of labour market forces or the corporate relocation decisions of their employer.

Migrants' views of quality of life

So far this chapter has provided evidence to show that societal, cultural, economic and technological forces have altered the relationship between people and place since the 1970s. The mechanics of this changed relationship have involved the emergence of new aspirations about desirable residential locations, new freedom to relocate the household as a result of technological, demographic and other factors considered above, and, thirdly, the repositioning of the residential environment as a consequence of new capital formations. This repositioning has operated to the benefit of some places over others and has resulted in a new set of attractive residential environments in specific places which are not necessarily in line with those offered in neighbouring places.

The outcome of these trends has been to make the analysis of the relationship between quality of life and migration complicated. Nevertheless, as the remainder of this chapter seeks to show from the authors' surveys, this relationship is comprehensible. There are several ways of investigating it. First, we compare the quality of life views of migrants with those of non-migrants in order to determine why migrants have a distinctive definition of quality of life. Second, we compare observed migration patterns with spatial variations in quality of life. We conclude the chapter by addressing the implications of the influence of quality of life on migration.

Every individual holds a personal view of what constitutes the 'ideal' quality of life. This is so not only because of individual variations in the importance attached to the key factors constituting a high quality of life (i.e. individual definitions of quality of life) but also because the very concept of quality is culture-specific, as well as being influenced by the conditions prevailing in different societies in specific, geographically and historically constrained contexts.

Since 1986 the Glasgow Quality of Life Group has sought to establish a new framework by which perceptual and so-called objective indicators of quality of life can be integrated to produce a ranking of British cities. The methodology is not discussed in great detail here since it has been reported adequately in a technical paper by Rogerson (1989; see also Rogerson et al., 1989). In essence, however, the method adopted consisted of examining 47 quality of life characteristics for a range of different cities (defined in terms of their local labour market areas) and local authority areas. These environmental, social and economic indicators were weighted in terms of their perceived importance to groups in the population, with the weightings being derived from specially commissioned national opinion surveys. The results of the surveys were used as major inputs to the analysis of quality of life and provided one of the most novel aspects of the methology developed by the authors for assessing quality of life in Britain (Findlay, Rogerson and Morris, 1988; Rogerson et al., 1989).

Table 3.3 Migrants' perceptions of quality of life dimensions

Dimension	% respondents indicating dimension important or very important		
	Migrants	Non-migrants	Signif
Pollution levels	88.2	89.5	
Health care provision	87.4	95.2	**
Violent crime levels	87.4	93.2	**
Cost of living	85.4	91.3	**
Non-violent crime levels	84.0	91.8	**
Shopping facilities	81.2	85.7	
Education provision	79.5	76.2	
Cost of owner-occupied housing	78.2	77.5	
Access to areas of scenic beauty	77.4	73.7	
Employment prospects	75.5	72.9	
Wage levels	75.2	75.4	
Travel to work time	68.8	66.0	
Leisure facilities	65.0	58.9	
Unemployment levels	60.7	68.5	*
Sports facilities	55.8	54.5	
Cost of private rented accommodation	45.1	48.4	
Quality of council housing	42.1	54.8	**
Access to council housing	38.3	51.5	**
Climate	34.6	58.5	**

Note: ** Chi-square coefficient significant at 0.01 level.
     * Coefficient significant at 0.05 level.
Source: Authors' survey.

All respondents in the stratified national samples were asked to rate each of the dimensions of quality of life on a scale ranging from 5 (indicating that the dimension was very important to the respondent) to 1 (indicating a feature of minimal influence), although each respondent also had the option of rating the dimension as 0 (indicating that it would not be considered by the respondent in the assessment of quality of life). A selection of the important dimensions of quality of life derived from the surveys are presented in Table 3.3, along with the results expressed in terms of the proportion of respondents who identified these criteria as important or very important. The table contrasts migrants' responses with those of non-migrants.

From the geographer's point of view, one of the interesting things about Table 3.3 is that it includes elements from both the physical and human environments, thus stressing the need for geographical skills in evaluating the total lived environment as perceived by the public, and not merely some limited aspects of places such as their economic or social fabric. Examination of the table also underscores the point that places cannot be rated meaningfully only in terms of economic indicators. Economic dimensions of life did matter to the British public in the late 1980s, but of primary importance were other dimensions of the quality of their living environments. In Table 3.3 only one directly economic variable emerges amongst the top five dimensions, while good health service provision and low crime rates head the list of criteria defining the ideal place to live.

What is particularly significant in the context of understanding population structures in places is, firstly, that the migrant population held a clearly defined view of quality of life and, secondly, that this view differed in several respect from that of non-migrants. Those who were involved in migration between 1984 and 1989 placed greatest emphasis on a pollution-free environment, low crime rates and good health provision, while by contrast only 1 in 3 migrants thought that climate was important in defining quality of life (Table 3.3). Employment-related dimensions of quality of life were neither rated top, nor were they ignored by respondents.

Chi-square analysis of migrant and non-migrant views of quality of life illustrate the distinctiveness of those who have migrated. This group place less emphasis on health provision, crime levels and cost of living as well as on public and private rented housing. Such a contrast is not surprising given the selective demographic and socio-professional nature of those groups most involved in the migration process.

Observed migration patterns and quality of life

However plausible the above arguments may be in making a case for quality of life as a factor in explaining a proportion of British migration, it does not necessarily mean that people's movements matched their quality of life aspirations. This section presents the results of two tests of the relationship between migration and quality of life at the aggregate level. The first examines a range of economic, social and environmental characteristics for a set of local authorities which attracted large net inflows of migrants during the 1980s in order to see whether these were offering a high quality of life, while the other presents the results of an analysis which cross-tabulates migration trends against quality of life for a set of local authorities characterised by large-scale net in- and out-migration.

Table 3.4 lists the fifteen local authority areas of non-metropolitan Britain which received the largest absolute net gains of migrants between 1981 and 1989 according to OPCS estimates (OPCS, 1990). For each one, information is given for seven separate indicators of aspects of quality of life, including not just the type of employment variables conventionally used in the modelling of inter-regional migration but also measures relating to housing, health provision and environment. It is immediately clear that these migration-attracting places each possess a range of quality of life advantages compared to the average and that these advantages have at least as much to do with non-economic characteristics as with labour market indicators. Indeed, several of the 15 places are characterised by either above-average unemployment levels (Tendring, Bournemouth) or below-average employment growth (Woodspring, Wokingham, New Forest, Huntingdon, East Devon) or a combination of the two (East Lindsey), whereas a clear majority are better off than average in terms of low house prices, good GP coverage and in particular low scores on derelict land and pollution (Table 3.4).

From the evidence of Table 3.4, it is possible to suggest that the areas making net migration gains need to be split into different categories – there are indeed a number of economic boom towns, but there are a larger number

Table 3.4   Social, economic and environmental indicators for non-metropolitan local authorities
experiencing large-scale net in-migration 1981−89

District (1)	Net in-migration 1981−89 000s	Employment growth (2) %	Unemployment (3) %	Housing costs (4) £000s	GPs (5)	Derelict land (6) %	Air pollution (7) ppm
Milton Keynes	+43.4	+8.4	4.0	66.8	0.59	0.00	37.0
Woodspring	+27.8	−0.9	4.2	61.2	0.68	0.14	30.5
Wokingham	+25.5	+2.1	1.5	87.8	0.50	0.00	28.8
Tendring	+24.6	+7.9	6.2	64.2	0.57	0.02	23.3
Northampton	+20.3	+13.6	4.3	64.3	0.55	0.65	29.0
New Forest	+19.7	+1.2	3.6	11.5	0.60	0.03	22.3
Arun	+19.7	+10.4	2.8	75.8	0.58	0.00	31.7
Wealden	+19.6	+4.0	1.7	85.0	0.59	0.00	29.6
Bournemouth	+19.6	+8.8	6.0	71.2	0.59	0.00	22.3
Teignbridge	+17.9	+7.1	3.8	57.2	0.61	0.01	17.7
Huntingdon	+17.4	+1.2	2.9	58.0	0.51	0.05	36.0
East Devon	+17.2	+1.4	3.2	88.1	0.61	0.01	25.5
Suffolk Coastal	+16.9	+13.3	2.3	54.5	0.67	0.01	27.5
East Lindsey	+16.2	−1.6	8.4	50.6	0.56	0.00	33.3
North Norfolk	+15.8	+10.5	4.3	61.7	0.55	0.01	25.5
Average of 145 Districts (8)	−	+4.0	4.7	70.0	0.58	0.50	34.2

Notes: (1) Non-metropolitan local authorities experiencing largest net in-migration, 1981−89;
source OPCS (1990); (2) Change in number employed, 1984−87; source: Census of Employment, 1987; (3) Proportion of workforce registered as unemployed, April, 1990; source: Dept
of Employment; (4) Average cost of 3-bedroomed semi-detached house, March 1990; source:
Estate Agents; (5) Number of General Practitioners per thousand residents, FPCs, 1989;
source: General Medical Services Statistics, Dept of Health; (6) Proportion of local authority
classified as derelict; source: Derelict Land Survey, 1988; (7) Concentration of sulphur
dioxide in atmosphere, average 1984−1988; source: Warren Springs Laboratory; (8) Average
based on study of 145 non-metropolitan local authorities of Great Britain with 100−200,000
inhabitants in 1989; source: Rogerson et al. (1990).

of local authorities whose chief attractive characteristic would seem to have
been their low pollution levels or their favourable level of service provision.
The inclusion of even limited environmental and social factors (Table 3.4)
enhance the level of explanation of migration flows in the 1980s. This
confirms that broader quality of life rather than just economic indicators
have to be considered if any meaningful explanation of contemporary migration
patterns is to be achieved.

The results of the second test are laid out in Table 3.5. This is based on a
study of the 145 local authority areas of non-metropolitan Britain which
had a population of between approximately 100,000 and 200,000 in 1989
(Rogerson et al., 1990). That study ranked all 145 places on the basis of
quality of life indicators. Here, the 26 places which gained over 10,000 net
migrants between 1981 and 1989 and the 8 places which lost over 10,000
people through net migration are classified according to whether they were in
the upper or lower half of the 145 local authorities on the quality of life

ranking. The weight of the evidence in Table 3.5 supports the contention that the areas with a high in-migration rate had a favourable quality of life rating, whilst those with a poorer quality of life experienced net migration losses. The rankings are based on the definition of a high quality living environment as seen by recent migrants included in the opinion survey (see Table 3.3). This finding is not surprising given the holistic nature of quality of life, but it does underscore the importance of environmental and social elements in influencing individual choices of migration destinations. Table 3.5 is compatible with other researchers' analyses of observed patterns of inter-regional migration.

It is obviously the case that only a minority of migrants moved to the most desirable places to live in terms of the Quality of Life Group's research findings. Table 3.5 does show, however, that the local authorities with the highest levels of in-migration and out-migration in the period 1981−1989 had, respectively, good and poor quality of life ratings. Despite offering a low quality of life to many of its residents, the South East inevitably continued to attrack selective in-migration, and not just of the upwardly mobile job-seeker; as Fielding (1989, p. 35) has commented, 'few bar the young could, or wanted to, gain access to the South East but, among those that did migrate to the South East, there were many who joined the urban underclass, and became part of the region's unemployed'. But for those who were in a position to choose where to live in a more unconstrained fashion, it would seem that the 1980s presented real evidence of many migrants opting for locations which could offer them a higher quality of life.

Of the 55,000 to 60,000 migrants who left the South East annually in the late 1980s the majority did not go to the highest quality areas in peripheral regions of northern Britain but to other areas such as the South West, East

Table **3.5** Quality of life rankings and migration in the 1980s: non-metropolitan local authorities (1)

Migration trends 1981−89	Ranking on basis of quality of life (2) (% of local authorities)		
	Above average rank	Below average rank	Total
Net in-migration greater than 10,000	19 (55.9)	7 (20.6)	26
Net out-migration greater than 10,000	0 (0.0)	8 (23.5)	8
Total	19	15	34

Notes:
(1) All non-metropolitan districts with in-migration or out-migration greater than 10,000 between 1981−89. On this definition, 34 of the 145 district councils in Rogerson et al. (1990) were included in this table; (2) Source: Rogerson et al (1990). All 145 non-metropolitan local authorities with populations between approximately 100,000 and 200,000 were included in the quality of life study. Rankings are based on the definition of quality of life given by migrants.

Anglia and the East Midlands where quality of life was nevertheless much superior to that of the major conurbations of the West Midlands and the South East. Fielding's work again suggests that many of the movers who left the South East in the 1980s were middle-aged professionals and managers and members of the so-called 'service class' who were opting to switch to self-employment and ownership of small businesses in regions which, for economic and personal reasons, were more attractive than the South East of England.

To summarise, the available evidence on patterns of quality of life and patterns of migration are conformable. The fact that so many respondents in the authors' national opinion survey (Table 3.1) stated that quality of life was their reason for migration would seem to be both plausible and with potential as a contributory explanation of many of the migration moves which were taking place.

Implications

At a time of low or zero rates of natural increase in population, the role of migration in affecting the local demographic structures of places has been enhanced greatly. As a result, local planners have increasingly found that certain changes in service demand over time have been strongly influenced by net gains or losses in migration. For example, selective out-migration from inner city areas has increased the proportion of poor elderly households there, while at the same time other areas have gained the 'rich' elderly through retirement migration. The selective effect of migration in this case is to produce contrasting demands on social and health services in the two types of localities affected. On the one hand, the inner city locality has experienced an increased burden in terms of the higher proportion of elderly seeking social services provided, and largely paid for, through the state system while, on the other hand, retirement migrants are more likely to enjoy private pension schemes and to have opted for private health insurance.

Greater migration flexibility in recent years has meant that the significance of spatial variations in quality of life has been reinforced. Even small unevennesses have been becoming significant in the attraction or repulsion of potential migrants at the local and regional scale. In place of the harsh divides in migration patterns experienced in the past, it would appear that a greater spatial variety of migration patterns is emerging, reflecting the local positioning of places within the national map of quality of life.

It is hard to imagine how these trends will evolve in the future, particularly in relation to societal and cultural influences on migration behaviour. The tendency for quality of life to become an increasingly significant determinant of migration behaviour will not be reversed in the foreseeable future, but the scale of the process seems less certain. It is highly probable that there will continue, for demographic and economic reasons, to be a considerable proportion of the population who will not play an active part in the labour force, and yet who will wish and be able to locate themselves in what they perceive to be desirable places to live.

Equally, as the requirements for retaining a well-trained workforce grows, the tendency for further migration is likely to be reinforced by employers increasingly subsidising the moves of those staff they transfer around the country (Salt, 1990). In addition, few would deny that in the 1990s further advances in telecommunications will permit a higher proportion of the population to be more independent of fixed office locations in the economic core of the country and thus provide greater possibilities of people choosing more freely and effectively where to live. One 1989 survey by the Henley Centre for Forecasting has suggested that up to 20 per cent of Britain's jobs could be done from home by 1995. Although the recent recession may delay the achievement of this level of home-based employment, teleworking has already contributed significantly to the growth of Britain's self-employed, so it is likely that more residents and businessmen, currently constrained by ties with London, will be able to live in those areas which offer them a high or higher quality of life.

Migrant choice of destinations appears increasingly to be moulded not only by greater availability of geographical information about the contrasting physical, social and economic character of places, but also by the way that this information has been presented and projected by place marketing organisations eager to attract both capital and human resources. The outcome of the migration process in the future will in part reflect the real, uneven distribution of quality of life opportunities, but will also reflect the perceived advantages of differing localities portrayed through place imagery which aims to position the portrayed locality next to the aspirations and requirements of the target population group.

It is a long time since Mabogunje (1970, p. 16) claimed that migration was a 'circular interdependent progressively complex and self-modifying system'. This description applies not only to the migration system itself but also to the impact of migration on the quality of life of different places. For migrants, there exists the opportunity for enhancing their quality of life through relocation, while for the non-migrant population the selective nature of the migration process inevitably produces cumulative positive or negative effects for the human resource base of the localities concerned. This operates through the cycle of events discussed above, altering through time both the demographic structure of local populations and the service-provision environment which is so important in defining quality of life.

Conclusion

Previous and continuing research has provided a reasonably clear picture of the selective characteristics of migration in Britain and the way in which it is changing the demographic structure of places across the country. Less clear has been the understanding of the motivations behind patterns of movement. This chapter has shown that, for many migrant groups, quality of life is an important factor in influencing migration decisions. Analysis of migration motivations indicates that people's conceptions of the social and environmental characteristics of places are just as important as economic dimensions in influencing population re-distribution. Whilst not all migrants will be able to relocate to the most desirable location, it is evident that for many of those who are able to migrate, the search for a higher quality of life is critical in determining their choice of local destination.

Moreover, for the reasons outlined above, it seems highly probable that the social, economic and cultural trends seen in the past decade or so will continue and, therefore, that migrants' aspirations will become increasingly influential in moulding future migration patterns. As well as having potentially very important implications for changing population profiles and for those involved in planning and providing for local areas, this prospect raises a set of major research challenges and opportunities. Chief among these is the basic, but very important task, of monitoring migration patterns as they continue to evolve over time. This work should go hand in hand with the sort of work described in this chapter, namely to obtain a clearer idea of the decision-making process involved in migration, including the degree to which population subgroups vary in their perceptions of quality of life and the extent to which they are free to follow such aspirations. Beyond this lies the task of articulating the central role of migration as a societal process linking people and place, building on the observation that quality of life, like other cultural traits, intersects with capital to produce new population patterns through migration.

The 1990s are particularly opportune for expanding the amount of research on migration in Britain. For one thing, the Working Party on Migration in Britain has provided a great deal of groundwork in terms of piecing together the evidence of migration in the 1980s and reviewing the current state of knowledge about processes involved and their impacts (Stillwell, Rees and Boden, 1992; Champion and Fielding, 1992; Champion and Stillwell, 1991). Secondly, British migration researchers are going to be better equipped with migration data in the foreseeable future than at any time in the past. The worth of the National Health Service Central Register for the continuous monitoring of migration has been proven, while the 1991 Census can not only provide a wealth of cross-sectional insights into migration over the previous twelve months but also, thanks to the OPCS Longitudinal Study, be used to examine the movements of individuals since both the 1981 and 1971 censuses. The Census also provides the wealth of local detail needed for relating migrants to places; for instance, in comparing the types of migrants with the characteristics of their origins and destinations. Finally, a number of special surveys, not least the British Household Panel Survey and two proposed new government surveys on households and 'working lives', are likely to offer much more detailed insights into individual migrations and the circumstances surrounding them. All this additional information should help towards the better understanding for the migration process and the more accurate anticipation of trends in population distribution and local population profiles.

To the authors of this chapter, there are two particularly important conclusions arising from the investigation of the relationship of migration and quality of life. The first is that, while the role of non-economic variables in the decision to migrate has been recognised (Rudzitis, 1991), this is an area which has received inadequate research and which demands much greater attention, given the way in which the importance of quality of life factors in influencing migration decisions has been increasing. Second, the individuality of place reflected in differing degrees of attraction to migrants has to be recognised as significant in explaining the complex pattern of migration which has been emerging in Britain. Whilst broad patterns of population redistribution, such as counterurbanisation, have attracted considerable research in the last few decades, the greater complexity in migration in terms of choice of possible destinations and new forces influencing migration decision-making point to the need for renewed research interest into migrant motivations and the role that population redistribution has on linking people and place.

4. THE DEMOGRAPHIC COMPONENT OF LOCAL GOVERNMENT FINANCE:

Impacts on resources, needs and budgets

Robert Bennett and Günter Krebs

Because the size, composition and rate of growth or decline of a population varies greatly between places, the local government units responsible for providing services in each area experience considerable differences in demand or need for services. They also have differing capacity to raise money to pay for local services as a result of differences in the size and composition of the local population. Demographic characteristics are one of the key determinants of local government service requirements: indeed they are so important that they usually determine over three-quarters of total local expenditure of local government in Britain. Similar relationships hold in most other countries. This should not be surprising: population is the main source of demand for most local government services.

In this chapter the relation of local government finance to local demography is examined in three parts: first, the relation between demographics and local finance through the three elements of local services, local resources and fiscal balances; second, how the need for services is defined among the population; third, how changing demographics, through its effect on changing needs, interacts with other features of local budgets. Where possible, comparisons are made between Britain and other countries, but the main examples are drawn from England and Wales.

Population and local finance

Services

The size of the local population and its composition is one of the most important factors influencing local authority expenditure requirements. It is so dominant because most local authority services are delivered either directly or indirectly to people. Direct service delivery to people includes education in schools and personal social services such as care of the aged. Many direct services also go to a household rather than to a person, e.g. refuse collection and disposal, planning and building regulation. There is also indirect service delivery to people. This includes street lighting, roads, leisure and parks services and libraries. In other cases the magnitude of local service demand is closely related to the size of population, for instance in relation to police and fire services, and roads.

The composition of the population as well as its absolute size is also an important influence on local service demand. The balance of old versus young people, for example, affects the relative demand for old people's homes, meals on wheels and particular mixes of leisure services compared with the demand for education, childcare and other mixes of leisure services. Similarly the balance of high- versus low-income groups of people affects the total extent of demand and its balance towards income-support services. The social composition also affects the disposition of the population to buy private services rather than use local public ones. And it also influences voting preferences and hence the political composition of the local council which in turn affects the choices by local government to provide different mixes of services in some areas compared to others.

There are, of course, also important fields of local government services that are less related to the local population. For example, the local business structure of an area has important influences on roads, planning and environmental services – influences which may differ considerably from population-related service requirements. In addition, there are important technical characteristics for some services that are affected by scale. Location is also important. For example, the costs of road salting, gritting and clearing of snow are much higher in some areas than others as a result of their climate. Wage rates and mode of delivery of services may vary between rural and urban areas, and between London and the South East and other areas. There is also the influence of non-resident population. In areas that are major focuses of tourist demand or second homes, local government has to provide a different mix of services, with different levels of demand over the year, compared to non-tourist areas.

Local resources

The size and composition of local populations is also a major influence on the resources available to local government to finance its services. Generally, local government has available to it four main sources of finance:

- Local taxes;
- Fees and charges;
- Grants from central government; and
- Capital finance through loans.

The balance between these different sources varies markedly between different countries, and between local government areas within the same country.

For *local taxes*, most countries have a mixture of several sources which combine local taxes on property with taxes on people's incomes and taxes on local consumption (through a sales tax, VAT, or tax on drinks and alcohol). Local government in Scandinavian countries has tended to emphasise income-related taxes; the USA tends to have emphasis at city level on sales and property taxes; the germanic countries (Germany, Switzerland, Austria and now much of Central-Eastern Europe) tend to emphasise local taxes on business, whilst Southern Europe and France tend to have a complex mix of local sales taxes, property taxes and social security taxes. Britain was unusual in that, until 1990, it had a very heavy reliance on a single tax on property. This dominance by property tax has been followed by many countries which were former British colonies. In Britain considerable reforms have occurred since 1990 to change the local revenue base. Generally it can be said that the higher the reliance on income-related taxes, the greater is the need for central government and other equalisation mechanisms that take account of variable population composition.

Fees and charges are a source of revenue for which local government acts more like the private sector in regulating demand and raising revenue through price setting. This approach cannot be applied to most social services, particularly for income support. But the potential for charging is considerable in the fields of leisure services, some environmental services, and at the margin of education and social supports. In all OECD countries the extent of charging for services is increasing as attempts are made to introduce a more direct relation of service provision to the level of demand.

Grants from central government are a major source of support to local government in most countries. Their existence is a response to two main factors: first, inequalities in local capacity to pay for services as a result of size and composition variation of the population; and second, an acceptance of a level of shared responsibility for many service fields where there is both a strong national as well as a local interest. As a result, all countries have complex mechanisms for transferring resources from central to local government which are usually based heavily on population sizes. In countries with a significant regional or state level of government, such as the USA or Germany, all three levels become involved in such grants. In the EC there has also been the important development of EC grants in some fields, notably in infrastructure and social fund activities.

Capital finance through loans is a means for government to finance expenditure today which will be paid for over a period of years into the future. For individuals, the major such loans are usually mortgages to purchase a house. For government, the mechanism is similar and is usually based on issuing bonds which pay to the lender a rate of interest in return for the use of their money. There are, however, a complex range of financial instruments available, although in most countries the market for local government borrowing is heavily regulated by central government. Capital finance is a major means of paying for large projects which may take a long period to build and for which the benefits will flow over many years; for example, a bridge, a large new road, a school building, a swimming pool or library. For such investments there is an obvious economic argument for spreading large repayment costs between a number of years. There is also an equity argument, that the costs of a large investment should be shared over both present and future local populations as users of the service. Unfortunately, capital finance is also a means of political expediency: politicians often find it easier to raise loans to pay for current services than to raise taxes. The local population, too, may prefer to pass on costs to future generations. As a result central government and audit bodies attempt to keep a rigid division between capital finance for capital investment projects, and recurrent finance for recurrent expenditures (on salaries and consumables). This protects both the total public sector borrowing requirement and prevents exploitation of future generations by present populations.

Britain has undergone a number of major changes in the local resource base in recent years. Over the fiscal years 1990/91 to 1992/93 a 'poll tax' (termed Community Charge) has been levied by local government. This was a flat rate on all adults in each local authority, and varied between areas from zero (e.g. in Wandsworth) to over £700 per head (e.g. in Haringey) with an average of just under £300; see, for instance, King (1989), Bramley, le Grand and Low, (1989), Gibson (1990). The poll tax replaced the 400-year-old system of local rates which were based on the rental value of property. Like rates the poll tax was the *sole source* of local taxation. From April 1993 a new tax is being introduced termed the Council Tax. This will return to the principle of a local property tax, but will be based on the capital value of property (i.e. normally this is the price at which it will sell). To simplify administration, house capital values will be assessed only within broad bands.

The demographic composition of different local authorities affects the tax resources under each of these taxes differently. For poll tax, the raw population numbers of age 18 and over was the crucial determinant. For a property tax the tax base is determined by relative demand for property (which determines its rental and capital values) as well as the absolute size of population measured through the number of households. Each change in local tax definition has occasioned further changes in central grants, in local use of fees and charges and in capital spending. As a result, the assessment of local financial issues in Britain has been rendered very complex.

Local fiscal balance

The relation between services and tax resources is a key concern for a balanced budget. But comparisons are complex. The way in which the revenues of a local government are raised depends on the mix of the revenues available to it and the extent to which each is used. Definitions are important in order to understand these differences. For example the revenue base depends on the precise revenue source discussed, e.g. a local income tax will differ considerably from a local property tax in the revenue it offers and the effect on different people; and a local tax differs greatly in its impact from fees and charges. The precise way in which the law defines the tax is also important, for instance the property tax in Britain based on rental value (the 'rates') is very different from the property tax (Council Tax) to be introduced in 1993 that is based on market values of property. The structure of the local economy is also important since this affects the total personal incomes from which revenues derive. This relationship is shown on the left hand side of Figure 4.1, which derives from the discussion by Smith (1976) and Bennett (1980). The population and economic base of an area provides the resource base, but the final level of available resources also depends upon the ability of local government to vary its tax rates, charges and other demands.

The resources available also depend on the 'need' for services and other spending. The definition of the concept of need is highly problematic since there is no easy way of distinguishing people's needs from their wants or desires (cf. Bennett, 1980, Chapter 5). However, for many purposes a limited concept of 'need' is employed which relates services to a common standard. In this sense, need is a measure of demand for resources. It derives from the socio-economic structure of the population of an area, as well as the needs of local businesses (for infrastructure, utilities, police and security services, etc.). Within a given local population a wide variety of 'client groups' exist which relate to the 'needs' for particular services. The cost per unit of providing each service varies considerably between areas mainly due to the size and density of settlements, climatic effects on the costs of maintaining roads and buildings, and labour and resource costs. The interaction of unit costs with client 'needs' and with variations in local authority decisions on how they wish to provide a service (particularly the effect of local political priorities) leads to the final set of expenditure decisions, shown on the right-hand side of Figure 4.1.

The balance between local revenues and expenditures leads to the overall local authority budget. Difference between revenues and expenditures leads to potential resources—needs surpluses or gaps. In the long term there is a legal requirement for budgets to balance so that gaps usually result in needs

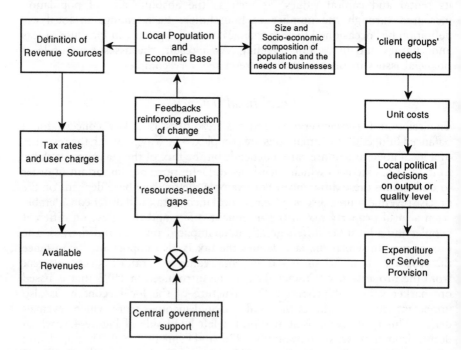

Fig. 4.1   Local government finance in relation to the size and composition of the population and economic base.

not being fulfilled because of insufficient financial resources. However, it is also possible for substantial deficits to be run, or for bankruptcies to occur, possibly giving rise to local 'fiscal crisis'. The extent of such 'crises' or 'gaps' also depends upon central government support for local resources, as shown in Figure 4.1. If 'gaps' do exist, they are likely to lead to strong feedback effects on the population and economic base. Even if fiscal 'gaps', as such, do not arise, any differences in public services in terms of their level, cost or quality can be expected to be an important influence on the decisions of people and businesses.

Migration of population or businesses is the most extreme response to these variations between places. The direction of migration flows will be to other areas with higher quality services at the same tax cost, or to areas with the same quality of services at a lower tax cost, or to areas with a mixture of the two. Such movement is termed 'fiscal migration'. More generally, there are effects on businesses or people that do not move but who suffer unequal service levels and/or unequal tax costs. This induces 'welfare inequalities'. Although definitions here are complex, the perceived form of inequalities will often reinforce, rather than diminish, existing social differences. For local businesses there will also be effects through variation in returns to different combinations of factor inputs and changed profitability between localities from what would have occurred in a free market: so-called 'distortions'.

Measuring the needs of the population

It is clear from the foregoing discussion that at the centre of debates about local government finance is the issue of need. Need has been defined pragmatically for most policy purposes as a measure of the standard services that an area would be expected to provide given its local characteristics.

Population factors play a dominant role in determining need measured in this way. In the estimates of need used by central government to allocate grants to local government in Britain since 1990, population accounts for over 50% of the government-determined Standard Spending Assessment (SSA). This represents an increase in importance of the specifically demographic aspects compared to earlier years. For example, in 1986/87 48.6 per cent of the standard need assessment was based on demographic information, with a further 14.2 per cent on physical features such as population density, 9.1 per cent on environmental and social criteria, and 24.9 per cent on 'special requirements' designed to take account of actual local expenditure choices and particular social service factors (ACC, 1985). Assessment of local needs of the population is a key aspect of both central government's Revenue Support Grant (RSG) allocation to local government, as well as local authorities' own management information. In various forms need assessment is responsible for influencing over £12 billion of central government grants and a further £10 billion of locally determined expenditure. In this section we discuss how need can be measured and how it has evolved in England and Wales in recent years.

Measuring need

The measurement of need is a complex task that is not amenable to solutions that offer unquestionable absolute indicators. Need is measured by indicators that seek often a generalised assessment of what the population requires. There have been three main approaches to this. *Individual indicators* can be used for clearly defined services with a readily specified client group. However, few individual indicators capture enough dimensions of need. Demographic data alone fail to measure social characteristics; technical data may fail to capture variations in volume. More commonly used are *composite indexes* in which a number of indicators are combined with a mixture of weighting factors. This method faces major dilemmas about the choice of indicators and weights, though Davies *et al.* (1971) and Davies, Banton and McMillan, (1974) have attempted to overcome these problems by use of principal component analysis. There are also considerable difficulties in measuring need independently of past levels of service provision.

The third method, which is preferable from many points of view, is a *representative need index*. This identifies a client group associated with each service category. This can be defined by multiple criteria which allow social and other factors to be taken into account. The weight applied to the client group measure is based on a unit costing. In addition, factors allowing for local discretionary differences in scope and quality of services can be added. The resulting need index is:

$$N^{kij} = \text{client group } ij \times \text{unit cost } ij$$

This defines need N for service category k to client group i in local authority j as a function of the number of clients and the unit costs of providing their services. A form of this approach, but with changing definitions over time, is that adopted in allocation of the central government's RSG since 1990/91, and the earlier system of Grant-Related Expenditure Assessments used from 1981/82.

Changing patterns of need

In the allocation of RSG, central government uses over 70 categories of service, to each of which a specific set of need indicators is attached. These need indicators attempt to measure the client group. To them is then applied a unit cost estimate. Most client group estimates are based on population information – usually for the expected eligible cohort. The result is a Standard Spending Assessment (SSA) for each local authority in England and Wales. There is a division of SSAs between the two tiers of county and district in non-metropolitan areas.

The estimates of SSAs are made each year for each area. But a major problem arises when we attempt to assess changing need over time. The definition of needs employed for central government RSG allocation has changed every year in major or minor ways under the block grant system used since 1981/82, and there is a major discontinuity in needs assessment between the old RSG demographic needs indicators used prior to 1981 and the block grant system used since 1981. Further changes occurred as a result of local poll tax in 1990/91. There is also a major break in continuity resulting from reorganisation of local government in 1974, a further change in the way RSG was allocated prior to 1974, considerable inconsistency in the use made by local authorities of SSAs or other information in their management, and major differences in the need assessment approaches used in England, Wales and Scotland. Although there has been a growth of comparative information (such as CIPFA *Local Government Comparative Statistics*), none of this provides long-term assessment of need over time.

Only by use of a consistent measure of need can the influence of demographic change be assessed. Since the SSAs change as the measure of need for each year it has never been possible easily to trace the development of needs over time, to identify the response of needs to changing demography, to assess relative rates of change in different areas, or to examine the relation of budgets to need in one period compared to another.

In order to overcome this difficulty a new estimate of needs has to be constructed. This is illustrated below by using a representative index to illustrate how needs to spend have evolved in England and Wales. The details of construction of this index are complex and the reader is referred to Bennett (1982, 1989b) and Offord (1987) for details.

The pattern of change in need to spend around the country using this index is shown in Figure 4.2. Only a small variation in national aggregate need occurs over the time period of this analysis, approximately 1 per cent, but the pattern of variation between local authorities is strongly variable. The more urban areas outside London have strong increases in need, particularly in their hinterland areas. Rural areas also show strong increases in need. In contrast, London, particularly inner London, and the central cities in the other conurbations show declines in needs. These changes are dominated by demographic influences. They reflect younger age cohorts in the rural and suburban areas. This in turn reflects the steady migration, since 1945, of younger people in their child-bearing years to suburbs and settlements beyond the main cities – a movement that reflects availability of suitable housing stock there and lack of private sector housing in inner cities, as well as people's preferences.

The major influence on changing need is the shift in the second wave of the post-war 'baby boom' generation. This large age cohort has recently left education and has entered the working age adult population. At the same time, the first wave of the 'baby boom' is still contributing a large element to the older working population (the 40s age cohort). The 5–16 age group is a particularly important cohort aspect of need since it forms the compulsory education age groups, and education accounts for approximately one-half of total local authority spending. Thus even small changes in the relative pro-portion of people in this age group can have major financial implications. The 18–21 age group also has important local expenditure implications through the demands for training and Further Education. These two education effects are the dominant influence on the changes in need in Figure 4.2.

The 12–16 age group represents the most expensive part of the education sector. This age group has reduced from a high in 1979 to a low in 1991. The

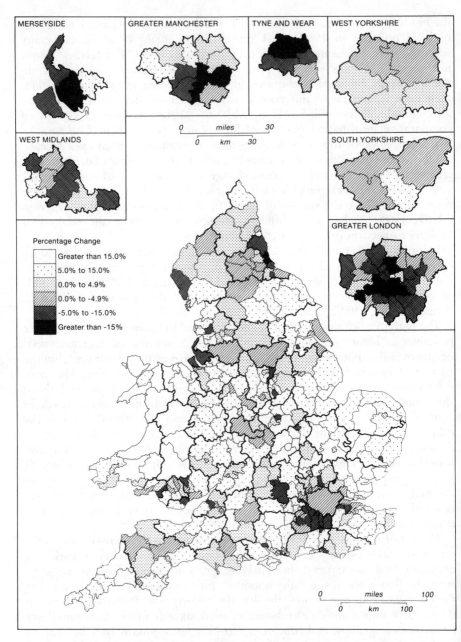

Fig. 4.2 Change in local need assessed using a representative index 1974−85 (*Source*: Bennett, 1989b. Shadings of metropolitan areas on main map are averages of their districts/boroughs).

reduction nationally was 30 per cent 1979–91. The change over Census periods is minus 25.7 per cent 1981–91 (source: OPCS estimates) or a reduction of 1 million in the number of 12–16 years olds 1981–91. However, this has followed a period of rapid increase of the 12–16s over 1971–81 of 14.7 per cent. Thus the 20 years 1971–91 have seen a population wave sweep through the schools and now into the adult age groups. Conversely the adult population has been rapidly expanding. The result will be an increase in the total adult working population until about 2011. The increase 1973–2011 is estimated to be approximately 9.8 per cent (Department of Employment estimates), whilst that for 1981–91 will be approximately 4.6 per cent; this represents an increase of 1.5 million people of working age.

The demographic impact on local budgets requirements is, therefore, seeing a major reduction in the most expensive sector (education) and a major increase in the potential revenue base (the adult population available to pay local taxes).

Demographic impact on local budgets

Standardised need assessment provides a means of assessing overall spending requirements but it does not describe the *total* local financial budget impact shown in Figure 4.1. We can go on to assess total budget impact by determining the relative importance of changes in need compared with other factors. A simple accounting methodology allows changes over time in key financial variables to be compared with a base year. This accounting framework uses the following equation:

$$
\begin{matrix}
\text{Actual tax rate} \\
\text{in given year}
\end{matrix}
=
\begin{matrix}
\text{tax rate in base year (1979)} \\
\pm \text{ changes in tax base} \\
\pm \text{ changes in central government grants} \\
\text{(mainly RSG)} \\
\pm \text{ changes in other income} \\
\pm \text{ changes in need to spend} \\
\pm \text{ changes in local expenditure decisions}
\end{matrix}
$$

Using this framework it can be seen that a specific local tax rate decision is the outcome of six main influences. First, there are the tax rates set in earlier years – these form a baseline of locally raised revenue from which only limited change is possible from one year to the next. Second, changes in the tax base are the resource available to be taxed. Areas experiencing population growth associated with new housing have an expanded tax base; areas experiencing population losses and housing clearance experience tax base decline. Third, there are changes in central government grants. These seek to reflect changing needs and changing numbers of the population, but they also reflect central government's political priorities. As we shall see, they are the main influence on local budgets in recent years. Fourth, changes in other income sources reflect decisions on fees and charges, revenue balances and other minor sources. New need forms the fifth source of influence on the local budget. It may change in a different manner to the tax base and central grants. Finally, changes in local expenditure decisions reflect the choice of local government to spend at a higher or lower level than previously.

Using this approach each variable in the above equation can be translated into an equivalent measure at constant prices to allow estimation of the components of change which contribute to equivalent increases (or decreases) from the tax rate in the base year. In the following discussion all changes are translated into a change in equivalent tax rates. As a result, changes in actual tax rates for a given year can be compared with the base year and the main components accounting for the observed changes can be identified. This methodology has been developed by Bennett and Krebs (1988).

The results of applying this methodology are shown in Table 4.1. This reports estimates for groups of areas, using the definitions of central city, hinterland and rural areas grouped into six categories (see Bennett, 1989b). The base year of analysis is 1979. In the table, changes in the main components of local authority budgets are translated to become changes in local tax rates over the period 1979 to 1988. In general, tax rates increase in all areas as a result of RSG (which has reduced), and other income (which has reduced). The increases have been offset only to a small extent by tax base increases, whilst need and expenditure have changed in a variable pattern in different areas.

In the table the tax rate (line one) represents the total change in real levels of local tax rates. There are real increases in all types of areas, but these are largest in central cities, hinterlands and rural areas. London has fairly stable tax rates over the period analysed. The rest of the table sums as a budget, i.e. all income items minus expenditure items equal the final tax rate at the top of the table.

A tax-base increase represents an increase in available resources and hence results in the potential to reduce local tax rates. Table 4.1 demonstrates that all areas experience tax base growth which can reduce local tax rates by an average of 28.6 pence in the pound. Central cities have the lowest rates of increase particularly in outer London, whereas the most rapid increases are in rural areas and city hinterlands. For most cases this is mainly a result of suburban and rural population growth with associated new housing increasing the local tax base, but business tax base growth resulting from the spread of business 'out of town' is also significant. The whole pattern evidences the process of decentralisation, or urban–rural shift, which is common to other Western countries.

Changes in central government RSG are one of the most important elements of change. Overall declines in RSG 1979−88 are responsible for a tax rate increase of 82.5 pence, but account for 105.6 pence increases in hinterlands and 100.8 pence in rural areas. In contrast, inner and outer London, despite experiencing large cuts in RSG in recent years, have lower increases in tax rates than elsewhere. RSG is, however, the largest overall element of change, and increases in its influence outwards from inner London.

Changes in other income are a small influence on budgets except in central cities and London. The major importance of other income in these areas reflects mainly increases in tax rates resulting from greater debt repayments. These in turn are the result of local government in the inner cities attempting

Table 4.1 Change in equivalent local tax rates, at 1991 prices, resulting from change in main budget elements 1979–88 in different types of area

Variable	Inner London	London area Outer London	Hinterland	Central cities	Rest of England & Wales Hinterlands	Rural areas	Average (401 areas)
Change in income							
tax rate	7.5	7.0	30.1	64.2	43.1	33.0	41.6
tax base	−26.0	−15.2	−32.6	−23.1	−37.4	−38.1	−28.6
RSG	29.0	51.3	79.2	81.8	105.6	100.8	82.5
other income	18.3	29.3	7.9	21.8	−0.3	−8.6	11.9
Change in expenditure							
need	−35.0	2.0	1.1	−5.3	19.1	30.6	4.8
new expenditure decisions	21.1	−60.3	−25.5	−11.0	−44.4	−51.7	−29.0

Note: Analysis excludes City of London and Westminster.
Tax rates are measured in pence, and set as equivalent to the domestic rates applying over the assessed period, i.e. the tax rate for Inner London is interpreted as a potential increase of 7.5 pence in the pound of rateable value. County-level information is disaggregated to district/borough level, using the district-county proportion of rateable value or other indicators.
Source: updated, simplified and recalculated from Bennett and Krebs, 1988, Chapter 4.

to maintain spending in the early 1980s by borrowing rather than raising taxes. In contrast, rural and suburban areas have been able to reduce local tax rates as a result of increases in other income – generally derived from receipts from sales of assets.

Total need to spend changes rather slowly over time, only 4.8 pence in the pound increase. But major differences are evident between areas. City cores experience the most rapid decline in needs as a result of total population loss and its changing age composition with a lower youth age cohort. In inner London this is one of the largest changes resulting from any factor, a potential reduction in tax rate of 35 pence in the pound. There is also a significant but smaller reduction in need to spend for central cities outside London, down by 5.3 pence in the pound. In contrast, the needs of outer London and its hinterland show a modest increase in need to spend. But most dramatic is the 19.1 pence in the pound increase in the need to spend in hinterlands outside London, and 30.6 pence in the pound in rural areas. The effects of the major changes in population numbers and composition noted in other chapters can be seen to induce dramatic contrasts between areas.

A further significant factor for many areas is the change in aggregate expenditure levels as a result of the exercise of local discretion to spend. This accounts for a 29.0 pence tax rate reduction overall, but there is a major increase in inner London of 21.1 pence. The largest reductions are in rural areas and outer London.

The contrasts in tax rate impacts on local government shown in Table 4.1 are strong between areas. In general, in city areas, declining population needs in inner cities have been balanced by increasing tax bases. But local expenditure decisions, as well as very large reductions in RSG, have reduced the extent to which tax rate reductions have been developed. In rural areas, and in hinterlands, a contrasted experience has occurred. Here increased population need has been balanced by more rapidly increasing tax base, but an even greater level of expenditure reduction has occurred largely as a result of experiencing very high reductions in grants. Local government in different places has, therefore, experienced major sources of instability from changing need as well as other factors such as change in tax base and central grants.

Conclusion

This chapter has examined the relation between demographic change and local government finance. It has documented some of the changes in needs and resources that have occurred in Britain as a result of changes in size and composition of the population. The examples used as illustrations have focused mainly on changes in needs rather than a cross-section at any single point in time. Changing age structure and population distribution has very significant impacts on local needs to spend, on central government grants and on the overall tax rates set locally. The example of England and Wales shows the expansion of need to spend in hinterlands and rural areas, with declining need to spend in central cities, particularly inner London. These changes are associated with population mobility and the age structure of the population (particularly the different concentrations of the school-age groups). Population also affects local tax base and central government grant levels.

But the analysis shows that many other factors as well as population affect local government financial decisions. These include local political choices, rising expectations that change the level of demand, elasticity of demand with income changes, and the influence of producer interests that prevent flexible adjustment of wages and numbers of employees to changing levels of absolute needs (see, for instance, Bennett, 1989a). The influence of some of these factors is assessed further by Pearson, Smith and White (1989) at a national level across major service categories. To a less detailed extent they are also included in the wider discussions of HM Treasury (1984) and OECD (1988).

A good example of the way in which these other factors intervene in the relationship between population change and local government spending is provided by school education, which is the single largest local authority service. The rapid decline in school-age population since the 1970s has been met by broadly constant expenditure levels (Audit Commission, 1986b). This is primarily because of the unwillingness of some local authorities to cut expenditure or inability to close schools because of political, parental and employee resistance, but there are other factors, too. Fixed costs cannot be cut at the same rate as marginal costs, so the scope for spending reductions will be low if education is running at close to optimum efficiency. Demographic changes may have wider effects on the rest of the economic system preventing economies occurring; for instance, by raising the general level of wages or increasing the level or quality of provision required. Lastly, the expansion of educational provision in the 1970s may not have completely met the increases in need that resulted from the boom in school-age children then, producing 'efficiency gains' that led to a response in the 1980s of maintaining spending to improve nominal quality.

The influences on local government finance are clearly large and complex. The population factor has enormous significance, particularly in a period when a major demographic change is occurring. Nevertheless, its influence has to be analysed alongside the influences that also affect decisions on expenditure allocation at all levels of government. The example of education should be salutary for interpretation of other areas. The next chapter goes into more detail about the role of demographic and other factors in this important and rapidly changing area of service provision.

5. POPULATION CHANGE AND EDUCATION:

School rolls and rationalisation before and after the 1988 Education Reform Act

Michael Bradford

Population change has a major impact on the education service of every country. Rapidly increasing or declining school rolls bring many problems that complicate the educational process. This chapter examines the effects of declining rolls in England. It highlights the local variations in changes to school rolls and examines the responses to them, both before and after the major set of reforms contained within the Education Reform Act of 1988. It considers the importance of local population change before and after these reforms on planning the provision of school places, because the overall response to demographic change is very different before and after the Act.

The number of children entering primary and secondary schools in England peaked in 1971 and 1977 respectively. After these dates the educational world was experiencing and responding to falling school rolls (Briault and Smith, 1980). The precise timing and extent of this general trend varied from one Local Education Authority (LEA) to another. At a more local scale, some places even bucked the trend. The response to this trend was a general policy of school closure and amalgamation but the ways in which it was implemented varied from authority to authority.

Although the trend continued for many schools beyond 1988, there was a major change in the response to falling school rolls that was indirectly produced by the 1988 Education Reform Act. For example, the introduction of parental preference and Grant Maintained Status, by which schools 'opted out' of LEA control, weakened the LEAs' power both to predict rolls and to respond to them. Thus the context in which demographic factors affect educational provision has changed and the local and national outcomes are different.

In general, the Act redistributed power in the educational service away from LEAs, downwards towards schools, in particular towards heads and governing bodies, and upwards towards central government and the Secretary of State for Education. Such mechanisms as the local management of schools (LMS), in giving more power to the schools, permitted greater variability at the local scale. Other innovations, such as the National Curriculum and its associated national testing, reduced the variability of provision, but the published results from this testing provided a means of comparing schools for parental choice and competition among schools. In many ways it is too early to assess the new balance of power among schools, LEAs and central government, but already there are numerous issues emerging that suggest the kinds of local variability that will emerge.

These changes in rolls and policies have taken place against a background of a relative increase in private education, which has been directly and indirectly encouraged by central government, either by design or accident. Since both the initial extent and the growth of the private share in educational provision have marked spatial variability, the local contexts of educational change in state provision are very different.

This chapter indicates the general demographic trends affecting the provision of school places and their spatial variability, demonstrates the different local responses to falling school rolls up to 1988, and then analyses the effects of the Education Reform Act on both rolls and responses. These issues are addressed in the light of a changing and locally variable relationship between state and private education. At the national level there seems to have been a blurring of the private/state divide in educational provision, resulting in the development of a private/state continuum. At the local level the picture is much more varied, with consequences for the extent of competition between schools, the degree of parental choice, and the potential to predict rolls and plan provision.

The role of demographic factors in affecting school rolls

Fluctuations in birth rates have a major impact on educational provision. This is well illustrated by the recent experience of England. The national fall in the birth rate by one-third between 1965 and 1977 had a dramatic effect on an educational service which, since the 1950s, had become accustomed to a context of continuing expansion. Perhaps it is not surprising that the service did not respond rapidly given this context and given the possibility that the initial fall might have been only a temporary downturn. Indeed there were many areas still experiencing increasing pressure on school places at a time of general decline. It is always easy, in retrospect, to criticise a failure to observe a structural change.

Primary schools

The number of applications to a primary school is influenced by net migration in the local area as well as the changing birth rate. During the 1970s most inner cities experienced net out-migration. Manchester illustrates this well. The metropolitan borough lost 18 per cent of its population between 1971 and 1981. The out-migrants were over-represented by women of child-bearing age. The out-migration and the fall in the birth rate combined to produce a 33 per cent fall in primary school enrolments over the decade and a fall of nearly 50 per cent to the trough of the later 1980s. Concealed within these LEA trends were even more dramatic declines in particular localities. The wards of Hulme and Beswick, in central Manchester, suffered falls in enrolment of over 60 per cent between 1972 and 1982. At the level of the individual school the variation can be even greater. A school's enrolment is affected not only by the changing number of children in its catchment area but also by the changing extent of its catchment area (for example, through the closure of a nearby school).

Since a school's capacity may expand or contract, at any one time it is usual to speak of its vacancy rate. It should be remembered that two schools may have the same decline in rolls but end up with different vacancy rates. One school may be overcrowded in the first place and remove some temporary classrooms once vacancies begin to occur, whereas another may have some vacancies initially and not adjust its supply of places in any way. In Manchester in 1982 some schools had vacancy rates of up to 75 per cent, while others were oversubscribed. Since many of the oversubscribed schools were in areas of overall enrolment decline, this indicates either that there were very local islands of increased population or that parental preference was already being exerted.

While inner city areas were experiencing rapidly falling rolls, some outer parts of very large cities were still having to cope with overcrowded schools and the unrest among parents over access to certain schools. Many schools amended the age at which they allowed pupils to enter primary education as one coping strategy. As well as delayed entry, LEAs also adopted the strategy of redirection, effectively changing catchment areas, even if later this proved no more than a temporary solution. Such problems and strategies were occurring in the outer suburbs of Greater Manchester in the mid-1970s. By the early 1980s, some of the same schools that had been overcrowded a few years earlier were being amalgamated.

Besides the contrasts in past experience between inner cities and outer suburbs, major challenges to primary school provision were being encountered in two other geographical contexts in the past two decades. One concerned the falling rolls and possible closure of rural primary schools, which formed one of a number of key services that were disappearing from villages and threatening the villages' survival. In rural areas schools were seen as foci of communities even more than in urban contexts. Again, though, there were considerable differences between places. During the 1970s some rural areas were gaining population. Although many in-migrant households had passed the child-rearing stage in the life cycle, in some places, such as the outer parts of the South East, there were enough in-migrant children to offset falling rolls of rural primary schools.

The other growth areas of the 1970s and 1980s were the new and expanded towns and some of the smaller cities such as Winchester and Macclesfield. In parts of these towns where new estates had recently been built, issues over primary schools were more concerned with oversubscription than falling rolls, because the provision of new services, both public and private, often lagged behind demand.

In short, the small size of primary schools and the limited extent of their catchment areas mean that their rolls highlight local variations in demographic change. Some may be experiencing the effect of increasing birth rates and net in-migration, while others may be influenced by declining birth rates and net out-migration. Such local variations in demographic effects on school rolls may even occur during a period of a dramatic fall in the national birth rate.

Secondary schools

The issue of falling rolls is more complex for secondary than primary schools because other than demographic processes become involved. Variations in the privatisation of education, reorganisation of 16–18 education, the structure of provision and the staying-on rate beyond the age of compulsory education produce different local contexts in which demographic factors operate. They mediate the effect of demographic changes and affect the issues involved in responses to them.

Some of the effects of very local variations in population change are felt less by secondary than primary schools because of their larger size and catchment areas. There are, however, still significant variations in the experiences of secondary schools both within and between regions. Figure 5.1 portrays the variation in the number of 11 year olds entering secondary schools in the North West and the South East at the LEA level. It shows the two elements of local differentiation: the varying timing of the peak year of entry and the variation in the extent of falling rolls. Both the timing and the extent of the demographic decline vary partly because of the effects of net migration and partly because of the social composition of areas. Greater falls occurred in areas where there was a higher proportion of social classes IV and V for whom the decline in birth rates was more pronounced. The peak year of entry was earlier (1975) in the central areas such as Liverpool, Salford and the Inner London Education Authority (ILEA), all areas of out-migration. It is later in the outer parts of the conurbations and Greater London and considerably later in LEAs such as Essex, E. Sussex, and Berkshire in the South East and Bolton, Oldham and especially St Helens in the North West. Although it seems that these North West LEAs experienced their decline later than those of the South East, it should be remembered that the spatial units are of different sizes. The figures for the larger shire counties reflect a much greater degree of averaging, or spatial smoothing, than do those for the smaller metropolitan boroughs. For example, the area around rapidly growing Reading in Berkshire may well have had a later peak than that for Berkshire as a whole.

Figure 5.1 also shows the drop in entry relative to the peak year. When interpreting the two noticeable trends, it should be remembered that many areas declined further after 1986. Overall, there is a greater decline in the North West than the South East. This is because the North West was the English region that lost most population during the 1970s and 1980s. Secondly the inner boroughs of the conurbations and inner London decline more than the outer areas, with Knowsley having less than half its peak entry by 1986. This is because the inner areas lost population at a greater rate than the outer ones.

There are a number of processes that affect the local contexts of secondary school provision and thus mediate the effects of demographic change. The

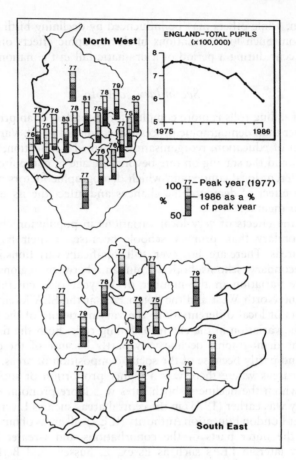

Fig. 5.1   Declining rolls: numbers entering secondary schools

actual decline in numbers entering secondary schools that are state maintained
was even greater than that shown in Figure 5.1, because from 1977 to 1986
relatively more 11 year olds were entering private education. This was particularly the case in Greater London where the impact of demographic decline
was accentuated by the absolute as well as the relative growth in the share of
private education (Bradford and Burdett, 1989a).

The impacts of demographic change have been different in LEAs that
reorganised their 16–18 education in advance or in anticipation of falling
rolls, compared to those that retained their initial organisation after becoming
comprehensive, that is all schools were planned to have sixth forms. For the
many secondary moderns that went comprehensive in the late 1960s and
early 1970s, it began to be clear by the mid to late 1970s that this goal might
not be attained. During the 1970s the number staying on beyond 16 did not
increase sufficiently to allow for all schools to have sixth forms, at least not
ones large enough to be considered financially viable. This meant that in
many areas they were changing their provision for 16–18s before falling

school rolls began to have an impact or in anticipation of it. Some LEAs favoured sixth-form colleges. Others decided to concentrate their 16–18s into a small number of their secondary schools. Some closed schools so that they might retain sixth forms in all their secondary schools. Some favoured a mixture of the above strategies while others had a mixed strategy imposed upon them.

The latter approach is well exemplified by the case of Manchester. Its LEA planned to have three sixth-form colleges, one in the north, one near the centre and one in the more prosperous south, where there was considerable opposition to the plan. The plan was based on encouraging equal access to sixth-form education across the borough. All such reorganisations have to receive the agreement of the Secretary of State for Education. Sir Keith Joseph's eventual decision to allow the northern and central colleges but not the southern one, justified on the grounds of the success of sixth forms in three southern schools, defeated the aims of the scheme and at the same time implied that other sixth forms in the borough were unsuccessful.

The approach in neighbouring Stockport was quite different. Whereas the teaching unions in Manchester had backed the original proposal fully, in Stockport they supported sixth forms in all the remaining secondary schools. In so doing, as in other authorities, they put off a greater rationalisation that in Stockport's case would create much local debate later in the 1980s when falling rolls began to have their effect. Thus some authorities reorganised their secondary schools before the impact of falling rolls because of difficulties involved with 16–18 education; others did not. The latter felt the impact of falling rolls much more severely.

The impacts of demographic change also depend on the structure of educational provision. When authorities went comprehensive, they created different structures of provision. While most had 11–18 schools, some created middle schools, 9–13, as in West Yorkshire, or in some cases, the US system of junior high schools, 11–14, as in Gateshead. Given similar demographic trends, authorities with middle schools face problems of falling rolls earlier than authorities with 11–18 schools.

Some authorities, usually Conservative-run ones, retain grammar schools, many of which, unlike comprehensives, are single sex. Falling rolls, in such authorities, have brought pressure to amalgamate two single-sex schools into one mixed grammar school, thus introducing the issue of separate or co-education into the debate about solutions to falling rolls. Such pressures affected actions later in the 1980s and early 1990s, as will be discussed later.

The staying-on rate, which itself may be influenced by the structure of provision, is the other factor which has interacted with falling rolls to affect the number of pupils in a school. It varies around the country with some areas, such as Hertfordshire, having a strong tradition of staying on, and others, such as Cleveland, finding that leaving at the end of compulsory education is very much more the norm.

Staying-on rates also vary over time. During the recession of 1979–82 staying-on rates generally rose. In some LEAs, such as Barnet, Gateshead, Harrow, Hounslow and Sefton, there was a ratchet effect and they remained at the higher levels as the economy improved. In others, such as the Isle of

Wight, South Tyneside, Sunderland and Solihull they dropped down again. Increasing staying-on rates at the same time as there was a demographic decline for the 16–18 cohort might lead to little change in sixth-form numbers; while a declining staying-on rate coinciding with a demographic fall might accentuate the need to reorganise 16–18 education. The effects of demographic changes are therefore mediated by changing staying-on rates. Thus, in addition to demographic processes, school rolls can be affected by economic processes and the traditions of the areas, which in themselves are related to employment structure and the social composition of the area.

For secondary schools, therefore, though demographic factors are clearly important, a much wider set of factors affect school rolls than at the primary level. This has been illustrated by the example of reorganisation at the 16–18 level, which in many cases began before demographic decline had its impact. Its eventual impact somewhat depended on the type of 16–18 organisation that was in place. Variation in the structure and organisation of provision also affected the timing of the effect of falling rolls and the type of issues that became embroiled in debates over solutions. Thus, even though most areas were experiencing demographic decline, its effects were mediated by the local educational, political, economic and social contexts.

Responses to falling rolls: the example of the pre-1988 period

There are two obvious responses to falling rolls: to have smaller classes and thus lower pupil per teacher ratios or to amalgamate or close schools. From an educational viewpoint there is much advocacy of smaller class sizes, especially in order to obtain the increased effectiveness of pupil centred learning, but some people object to this style of learning and dispute the evidence for increased effectiveness. On the other hand, it is argued that larger schools can offer greater opportunities for specialisation and choice, both for the pupil and the teacher. Clearly these two lines of argument are not mutually exclusive. There can be large schools with smaller class sizes. The debate usually revolves around the survival or closure of smaller schools.

Educational arguments, however, did not dominate the debate during the 'baby boom' period. Financial and economic arguments were uppermost throughout the late 1970s and the whole of the 1980s, as first the Labour government began to cut public expenditure after the IMF loan in 1976 and then the Conservative governments focused their attacks on public expenditure on the control of local authority spending. Cost cutting was one of the 'six Cs' that characterised education in the 1980s (Bradford, 1989) (the others being, contraction, careerism, commodification, choice and community control). In their emphasis on cost cutting, central government was aided by academic work on falling rolls (Bailey, 1982; Briault and Smith, 1980) which also emphasised economic rather than educational arguments. Since the costs of education are usually measured in cost per pupil and teachers are by far the highest proportion of total costs, the response of smaller classes would involve much higher apparent costs. The response of closure would, however, involve numerous hidden costs, such as the costs of not using a building, if it is not possible to dispose of it, or the loss of a community asset, if it is sold.

There are also the costs of the public debate about a closure and the possible costs of loss of community involvement as a consequence of closure. At this time the guidelines from national government were clear (DES, 1977, 1978, 1979). LEAs were to rationalise their provision in the light of falling rolls by closure or amalgamation. Resources had to be used fully and efficiently. The economic argument was dominant in the late 1970s and particularly the 1980s. It came even more to the fore because many of the authorities that were hit first by declining secondary school rolls were also ones affected most by reductions in central government Rate Support Grant. This partly reflected their loss in population but it was also due to the reformulation of the grant in 1981 which aimed at encouraging reduced spending (see Chapter 4). Inner-city authorities with their many social problems were especially high spenders, particularly on education. Penalties for over-spending and rate capping further restricted authorities room for manoeuvre. So even if some authorities did not wish to follow the closure route, eventually they had little option.

In some areas, the redrawing of catchment areas or the reallocation of feeder primary schools looked as though they might solve the problem of falling rolls in the short term, particularly where there were differential rates of decline in contiguous areas. Such reallocations of children were often inwards towards the centre of cities. These solutions, however, can arouse as much parental opposition as closure. Parents argue that they choose their house on the basis of the school catchment area in which it is located or the primary school on the basis of the secondary school to which it feeds its pupils. The comparative reputation of schools is the overt reason for such opposition. Often the covert reasoning involves the social composition of the alternative school and the desire not to have their children educated alongside children from different, and implicitly worse, social backgrounds.

Amalgamation and closure, however, were eventually the main responses chosen to deal with falling rolls over the period. As indicated earlier, LEAs responded slowly. They were used to an expanding system; a downturn in the birth rate presented a solution to overcrowding that had been the problem in some areas; and there was always the thought that birth rates might turn up again soon. For those advocating closure, the failure of LEAs to respond in relation to primary schools is understandable, because there is much less forewarning and the situation may change quite rapidly. For secondary schools, there is much less excuse. However, while the 1960s' consensus that an increasing investment in education per capita was an indispensable condition for the achievement of greater equality of opportunity and national economic prosperity (Brown and Ferguson, 1982) had collapsed at the national level, it still had supporters at the local level. Some LEAs tried to avoid closure because of its social impact or because of political unpopularity in what might be marginal wards. Others moved faster in response to central govern-ment's economic arguments and financial constraints. Most authorities proceeded on a piecemeal basis, amalgamating schools here, closing a school there. The criteria for the selection of these schools were as likely to be political as educational, a school with a weak or retiring head being a prime target for closure. Other LEAs drew up plans for the whole authority and presented it to the public for comment.

Manchester did this for its primary schools in 1982, advocating closures in areas proportionately to their demographic decline. Liz Bondi's analysis of the plan and the responses of local communities (Bondi, 1986, 1988) emphasises the dominance of economic rather than educational or social criteria in the rationalisation of schools and demonstrates the significant difference between local areas in their ability to oppose proposed solutions. It is not surprising that middle-income groups and/or professionally led groups were most success- ful, but even in poorer areas professional male-led groups were successful. It is difficult to disentangle whether it was the gender or the professional occupation of leaders that was significant, but professional male-led groups may have been most successful, because it was mainly male professional officers with whom they were dealing.

The tactics employed were also important in the Manchester case. The combination of what Bondi terms the 'factual' and 'political' approaches was most successful. Those only employing the 'public opinion formation' approach were unsuccessful. This encouraged people to oppose the plans through media-catching tactics, such as demonstrations and rallies. The factual approach used formal channels of consultation, presenting researched reports as well as letters of protest. The political approach involved the successful eliciting of support of local councillors. Those low income areas that were led by professional males and used the factual and political approaches managed to obtain reprieves for their local schools. Closure, then, did not have to do with demography alone but also with the leadership and tactics adopted by local groups.

Stockport, unlike its neighbour Manchester, had retained sixth forms in all its secondary schools earlier in the 1980s, as noted above. When demographic decline began to affect its secondary schools, it also moved towards the sixth- form college idea. As in Manchester, this aroused much opposition, especially in two of the three major areas into which it was divided for administrative reasons when local government was reorganised in 1974. The solution was a phased introduction of colleges: the first in the north where the problem of sixth forms being too small was most acute and where opposition was least; the second in the east and the last in the west where opposition was greatest and the problem least pressing. This phased solution appeased parents with pupils in the secondary schools at the time in the east and west areas because their children would not be affected. At the time it was regarded as a very clever political move. Later it was to be seen in a different light, because the advent of the 1988 Education Reform Act changed the context of educational planning so much.

In summary, the responses of authorities to demographic decline were therefore delayed, in some places much longer than others. The timing and extent of decline have been shown to vary considerably, and the responses have been shown to vary in time too, producing a very complex set of lagged outcomes. The responses were most often piecemeal, producing even greater local variation. Where there was an overall plan, financial arguments dominated the solution but some local areas, even relatively poor ones, were able to prevent the closure suggested by demographic factors through their leadership and tactics. Thus there were various responses at different times to demographic decline, accentuating in their outcomes the local variation in decline that has already been noted.

The 1988 Education Reform Act

This Act was the culmination of a number of changes that the Conservative governments of the 1980s introduced, many of them designed to extend market forces and commercial practices into the education service. Here it is proposed to discuss only those that affected rolls and influenced the response to demographic change. The changes introduced by the Act have effectively stopped LEAs responding to demographic decline and made it much more difficult to plan educational provision because they have created so much uncertainty in the education system.

Parental preference

The prime change that can potentially affect rolls is the Act's extension of parental preference. The rhetoric terms this 'parental choice', but in reality it is a set of preferences that is constrained by the capacities of the schools, as measured by places offered in 1979. Not everyone therefore can obtain their choice. However, it does seem to bring to an end the two major allocation procedures that have been applied to most secondary schools until 1988, namely catchment areas (or somewhat softer versions of them) and feeder primary schools. The 1980 Education Act had already allowed an expression of preference and the right of appeal, so the change is not as dramatic as the politicians make it seem. The rhetoric, however, does have its effect and it switches the emphasis to the parents and to the schools and away from the LEAs, although they still adjudicate on appeals.

The major difference from the situation up to 1988 is that parental preferences in the aggregate are to be used as expressions of the excellence of the schools. Resources will then follow the pupils and supposed excellence will be supported, while unpopular schools, facing declining resources as well as pupils, will have to improve in order to survive. If they do not attract more pupils they will close. In this way market forces will create competition between schools which supposedly will in itself promote excellence and improved standards. Suffice it to say that these are no more than assumptions.

In order to have some ways that parents can easily judge schools, the national tests and GCSE results associated with the National Curriculum are to be published, along with other so-called performance indicators such as the extent of truancy in the schools. The published results act as the currency of the new market system. They are also supposed to be a means of accountability for the investment of public funds. The results are to be published without any adjustments for the social composition of the school or the prior attainment of the intake. These omissions make the resulting school league tables uninterpretable in terms of the schools' effectiveness (Bradford, 1990). So too does the failure to adjust for the social geography of catchment areas

(Bradford, 1991), since the local area where pupils live, as well as their home background, affects their attainment (Moulden and Bradford, 1984). Since attainment is strongly influenced by social class, the unadjusted league tables will tell parents more about the social composition of schools than about the varying possibilities of their children realising their potential within the schools. Yet, as mentioned earlier, when discussing the redrawing of catchment areas, parents are often as interested in the type of children with whom their children are educated as with the quality of the education.

The longer-term impact of these changes on enrolment has still to be assessed, but the evidence from Scotland, where parental preference has been operating since the early 1980s, provides some indication of the likely effects in England and Wales. Compared with what would have been expected given the old catchment areas, most schools have been either net gainers or net losers (Raab and Adler, 1988). There are very few where the number coming from outside the old catchment area is balanced by a similar number moving out from the catchment area to other schools. Most of the net gainers are in middle-class areas, while most of the net losers are in council estates or lower-income areas. The future for schools in England and Wales that are located in peripheral council estates or inner cities where the children are within easy travelling distance of schools in somewhat better-off areas seem rather bleak. Unless they already have very good reputations, the publication of unadjusted exam and test results may well add to their difficulties of attracting pupils. The social composition of the schools will in many cases produce poorer results than schools in more middle-class areas, even though the 'value added' to pupils while they attend the schools may well be greater (Bradford, 1991). In short, they may be more effective schools, but unadjusted exam results will conceal this fact and their unpopularity may cause them to close. The 1988 Act therefore introduces a market forces model of producing closure in contrast to one based on local demographic change.

Two obvious consequences follow from these changes as far as school rolls are concerned. First, it is more difficult to estimate the future demand for a school. Second, in order to allow choice within the system as a whole there must be a considerable number of surplus places.

Before parental preference was introduced, LEAs estimated the future rolls of primary schools by examining the registration of births in the area and estimating net migration of households with children of pre-school or school age. For secondary schools, the numbers in feeder primary schools along with estimated net migration provided a relatively easy method of predicting rolls. In both cases, some estimate would be made of the loss to the private sector, which in most places during the 1980s was an increasing number (Bradford and Burdett, 1989b). Since the 1988 Act, such estimates have also had to build in the effects of parental preference. In rural areas there is relatively little difficulty because of the lower density of population and the greater distances which pupils would have to travel to go to different schools. In urban areas, by contrast, the likelihood of children travelling beyond former catchment areas is considerable. Here, therefore, either some estimate of the schools' reputations is also needed or, failing that, perhaps the relative social compositions of the schools may act as a surrogate measure in order to estimate future applications. For schools where oversubscription is predicted, it might be necessary to include some feedback measure for future applications which acts to dampen them down. This would take into consideration parents who, knowing that competition is great, might not apply to the oversubscribed schools, and put a preference for one which their children are bound to obtain if it is put first, but not if it is placed second.

As regards the need for a considerable number of surplus places to make the new system work, the government in 1992 suggested that capital would be available for popular schools to expand, so the 1979 capacity constraint may prove to be a temporary limit. This constraint has already caused some problems because the demands on school space have changed since 1979, with computer laboratories, for example, occupying classrooms that in 1979 would have accommodated more pupils. This will have been particularly the case for schools that adopted the Technical and Vocational Educational Initiative (TVEI) at a time when it involved a considerable amount of extra resources. Taking pupils up to their 1979 capacity has meant that many such schools are now overcrowded.

The provision of more places in popular schools may provide the surplus places needed in the system for parental choice to be meaningful. On the other hand, it may be used as an extra argument to close unpopular schools. There are two possible scenarios which highlight the effects of the 1988 changes. First, market forces could result in the closure of schools with large numbers of surplus places and insufficient resources to continue to provide the National Curriculum. If this occurs, there will be fewer schools from which to choose and the idea of parental choice will have been sacrificed. Perhaps more importantly, the schools that are likely to suffer such closure will be serving areas where there are many socially disadvantaged children. They are least able to afford any extra travel cost, so adding to their disadvantage. So school closure after 1988 is more likely to occur in areas of social disadvantage than of demographic decline. There might be social tensions from parents with pupils in the schools to which they may be switched. Such indirect local political pressure may present a greater chance of keeping the schools open than any pressure from the socially disadvantaged area itself.

The other scenario is that many more surplus places are allowed to remain in the school system so that some choice may occur, in which case there will have been a complete reversal from the earlier 1980s and late 1970s when efficient use of resources was all-important and the removal of surplus places by closure was central government policy. In this scenario it is not the location of closures that will be different but the amount of them.

Other innovations

The idea of Grant Maintained Status (GMS) or schools that opt out of LEA control has made the second scenario more likely in the shorter term. Any attempt by LEAs to plan provision in the efficient manner that was demanded in the earlier 1980s by proposing closures and amalgamations can now be met by the targeted schools threatening to opt out. The first set of Grant Maintained schools included many that were or had been under threat of closure. Their survival in a different form has already indirectly retained surplus places in the system as well as changed the system in other ways, as will be discussed in more detail below.

The introduction of City Technology Colleges (CTCs) – launched in 1986 and reinforced by the 1988 Act – has also created problems for efficient planning of resources for LEAs. The original idea was to target inner city areas and to restrict the size of catchment areas so that they contained around 5,000 secondary age pupils. A CTC of 1,000 pupils would then draw about a fifth of the pupils away from five existing inner city schools of equivalent size. Since by 1986 some LEAs had reorganised their provision, they were not too enamoured with an innovation that would cause further disruption and demand another round of rationalisation. Indeed it was an innovation that created schools outside LEA control and, in that sense, set the scene for Grant Maintained schools. Not long after the announcement of CTCs, national government enlarged their intended catchment areas so as to avoid having too great an effect on any one school's enrolment. This meant, of course, that they would no longer be as closely targeted on the inner city as originally planned and that pupils from further afield could attend them. Thus one goal was sacrificed somewhat for another.

CTCs, however, have so far not developed as rapidly as expected. They were supposed to be financed partly by national government and partly by private business, but there has been a marked lack of support from the private sector. As a result, the planned number of 20 by 1990 did not emerge; indeed there are likely to be only 15 by 1993, most of these supported much more by national government than private money. Therefore few LEAs have been affected and those that have been affected have usually not opposed the plans – in contrast to the many authorities which put planning controls and other difficulties in the way of any proposals.

Where they exist, CTCs, especially those in conurbations, present problems for the rolls of schools not only in the LEAs in which they are located but also in neighbouring authorities. With their wider catchment areas, they can easily tap demand across LEA boundaries. For example, Kingshurst in Solihull (Walford, 1991), the first CTC to open, has competed with schools in neighbouring Birmingham, the nearest of which was an early TVEI school that was very oriented to technology. The newness, the considerable resources and the promotion of the CTCs make them difficult to compete against in the era of open enrolment. Thus future planning of provision in neighbouring authorities can be as difficult as such planning within the authority in which the CTC is situated.

To summarise, it is clear that recent changes in educational provision have had important implications for school enrolment and made any planning or rationalisation of school provision much more difficult. Various parts of the 1988 Education Act, parental preference and open enrolment, GMS and CTCs have all made it much more difficult to plan future provision in the financially efficient way that was advocated by national government before 1988. While LMS, another part of the 1988 Act, might make the running of individual schools more efficient in some ways, there seems to be a loss of efficiency at the scale of the whole local system.

Grant Maintained schools: the effects of 'opting out'

It is worth considering the Grant Maintained (GM) school, one innovation of the 1988 Act, in greater detail to see the effects that it has had on enrolment and the functioning of the educational system, to examine the issues that it raises and to demonstrate the increased local variation in educational provision that is currently offered to children.

The author has carried out research on the first 44 GM schools and the LEAs from which they have opted out (Bradford and Hall, 1992). The 44 include proportionally more grammar schools than would be expected from the national set of schools and many, including several of the grammar schools, were or have been under threat of closure. In some cases the threatened closure was due to falling rolls within the LEA as a whole. Sometimes the school itself was rather small and losing numbers. In some cases, the single-sex grammar schools were under threat of amalgamation to be co-educational. In others, the grammar status was under threat. One school was not under threat at the time, because a recent decision had been made to retain it. It had been under threat for 20 years, however, and once GMS emerged, its governors grabbed the opportunity of removing it from any future threat. The great variety of circumstances under which the first set of schools took on GMS is illustrated by two comprehensive schools in Milton Keynes which opted out to avoid the pressure from their LEA to become grammar schools in line with the rest of the LEA's system. Despite this overall variety, there is little doubt that GMS has allowed many schools to remain open which otherwise would have been closed by LEAs as part of their response to falling rolls.

The responses of LEAs to 'opting out' and their relations with GM schools varied considerably. One LEA seemed positively to support GMS, saying that 'it is the policy of the members that GM schools should be treated as part of the county's provision', and that 'the LEA sees GM schools as one element of choice in education'. Another changed its view rapidly when it feared a domino effect within its county and wanted to retain the GM schools within the set of schools for which it provided services. A few LEAs regarded themselves as neutral. Most had campaigned against the opting out. Some still rather reluctantly provided services afterwards for the sake of their own remaining schools. They feared that the loss of the GM schools would result in the loss of economies of scale of service provision. By far the majority of LEAs campaigned against and offered none but statutory services to the schools. Relations between these schools and the LEAs were 'frosty', a word used many times in the interviews with both schools and LEAs. So it is not surprising that many GM schools do not inform LEAs about the children to whom they have offered places. Since many parents hold places at more than one GM school as well as an LEA school, it becomes very difficult to know which pupils will be attending which school at the start of a new year. Greater uncertainty is introduced into the process of anticipating the size of school rolls.

Co-operation on enrolment between GM schools and neighbouring LEA schools has been made even less likely by the divisive nature of 'opting out' which in many areas has led to no staff or pupil contact between GM and LEA schools. It has also led to even more intensive marketing by LEA schools to counter the promotional activities of GM schools. Such competitive promotion introduces even more uncertainty into the process of establishing rolls.

Therefore the experiences surrounding the first set of GM schools have been very varied. They have created greater local variability in educational provision, produced a more costly system by remaining open and, at the same time, because of the competition and division, they have made it very difficult to establish the future rolls of LEA schools.

Issues raised by GMS

A number of other issues are raised by GMS that impact on school rolls: various forms of selection; the relative roles of governing bodies and heads; resourcing; and the future of LEAs. They all contribute towards greater difficulty in the rational planning of educational provision.

Selection does not necessarily affect the total enrolment in other schools; rather, it affects the kinds of pupils enrolling. The admission procedures of GM schools were not supposed to change in the five years after their change of status, but the Secretary of State in 1991 began to soften this condition. There is now pressure from some schools to change their procedures. Since there are already many grammar schools within the set of GM schools, it is feared by supporters of comprehensive education that GMS is a backdoor route for the return of selection by ability. It is also feared that, once accepted for GM schools, such procedures would spread to LEA schools, and what is supposed to be 'parental choice', in effect, would become 'school choice'.

The issue of selection has also arisen in relation to race and ethnic groups, most notably the case of the Stratford GM school in East London. GMS may be seen as an indirect way to obtain state support for separate Muslim education, when such state support has been refused within the LEA system (in spite of support for the separate education of other religious groups). More generally this raises the issue of the chance of influencing the enrolment of ethnic groups in schools since the 1988 Act. There are no longer the means to operate a desegregration programme.

The 1992 Stratford dispute between some of its Asian governors and the Head also brought to the fore the question of the relative powers of governing bodies and heads and which body should act as an arbiter in such disputes between them. In schools controlled by the LEA, responsibility lies with the LEA whose officers have local knowledge, whereas in GM schools it seems as though it is the Secretary of State for Education who must perform the role. This highlights the centralisation associated with GMS as well as revealing the uncertainty about the relocation of powers and the potential for a few people as governors to decide enrolment in schools.

GMS has also stimulated further discussion on the resourcing of schools. The cost per pupil varies considerably among LEAs and some schools opted out because they were being under-resourced. Schools that opted out early on have benefited financially by being financed directly from national government but it is unlikely that future GM schools will continue to receive such advantages, given the pressure on public expenditure. After the general election result of 1992 and therefore the continuation of GMS, some LEAs considered the idea of all their schools opting out. Purportedly this was to avoid the divisiveness that has occurred elsewhere where some schools have opted out. It would also mean that the LEA would continue to provide services for all the schools, the uncertainty of the piecemeal opting out process would be averted and there would be a body co-ordinating enrolment by parental preference.

Wholesale opting out would at least retain a role for the LEA. GMS has raised the question of the relative roles of local and central government. GM schools can obtain their services privately rather than from their former LEA or, as one school suggested, it could obtain them from another LEA which was more willing to deal with GM schools (Bradford and Hall, 1992). Some GM schools have grouped together so that they can benefit from some economies of scale of ordering without using LEAs. This has suggested to some that LEAs should disappear and be replaced by regional bodies that could co-ordinate and provide some services. This would replace a demo-cratically accountable local body with one similar to the water authorities before privatisation, less accountable and more distant. Such suggestions are in order to avoid complete centralisation. Already some GM schools are concerned that the freedom of manoeuvre that they have gained from opting out of LEA control is being offset by a very distant bureaucracy in London which will not cope if GM schools multiply in number. The possible demise of the LEA removes the obvious local body that rationalises school provision in the light of changing rolls.

All this uncertainty about the number of GM schools and admissions policy, as well as about whether LEAs may continue to exist and if so in what form, makes any rational planning of future provision extremely diffi-cult. This is illustrated by Stockport LEA which, as discussed above, had implemented a phased introduction of sixth-form colleges. As the final phase occurred, the Conservative government decided to remove further education and sixth-form colleges from the control of LEAs. Now some of the schools that are losing their sixth forms in the west area are considering opting out so that they can retain them. If they do so, and if they are successful in attracting sixth formers, the viability of the new sixth-form college is called into question. In all these changes the idea of using resources efficiently has been overlooked. The pre-1988 system, whether intentionally or uninten-tionally, has been superseded.

This change has now been recognised in the 1992 White Paper on Education which advocates the establishment of a central body to rationalise 'surplus' places (Department for Education, 1992). The White Paper, however, does not tackle the tension between using resources efficiently and allowing a sufficient number of 'surplus' places in the system to permit parental preference to operate effectively.

Local variations in the evolving private/state continuum

The changed state of the education system in England can be summed up at the national level as a blurring of the private/state divide. This has made it more difficult to plan for state school enrolment. In some areas, though, there is much more difficulty than others, because the development of the continuum has been very patchy.

During the period 1976 to 1979 there was a clear distinction between private and state schools. The former Direct Grant Schools, which had blurred the distinction before that time, had been abolished by the incoming Labour government. These had educated both fee paying pupils and very bright ones who had taken entrance exams to obtain free schooling. Two-thirds of these schools became private.

When the Conservatives entered office in 1979 they introduced the Assisted Places Scheme (Edwards, Fitz and Whitty, 1989) which paid for very bright pupils to go to selected private schools, a form of state subsidy to private education. The later introduction of CTCs which involved private finance and GM schools further blurred the divide. The latter involved not only removal from LEA control — a form of deregulation — but also the transfer of assets (land and buildings) from the LEA to the schools. The resulting private/state continuum is shown in Figure 5.2.

While this continuum is characteristic of the national situation, however, there is much variation at the local level. This variation may be demonstrated by examining the introduction of GM schools alongside changes in private education. The spatial distribution of GM schools is very uneven, with many LEA areas possessing no GM schools as yet and some having many.

This emerging diversity across the country can be illustrated by an analysis of the changing share of private education for the extreme cases of county LEAs that had many GM schools and those with none declared in the first 100. In Figure 5.3 these two groups are further subdivided according to whether the counties are more or less oriented to private education compared to their regional average. The resulting four sets of counties clearly demonstrate that the national picture of a private/state continuum is far from representative of all local situations.

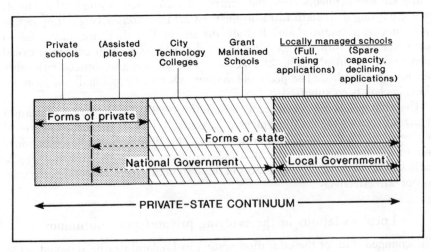

Fig. 5.2   The private/state continuum in educational provision

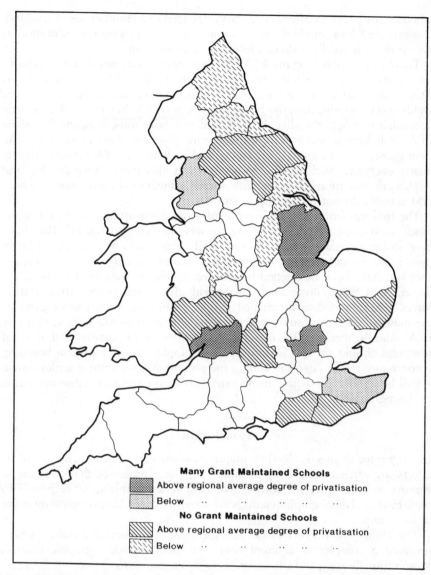

Fig. 5.3   The incidence of Grant Maintained Schools: the extreme cases

The first set of counties (Herts, Lincolnshire and Gloucester) have a strongly developing continuum, with GM schools extending the penetration of the 'market' already made by the private system. For the second set (Lancashire and Kent) there is a continuum, but GM schools could be regarded as filling the relative gap in private provison and bringing the counties towards the regional average of private penetration of the 'market'. The other two sets of counties have no GM schools. In the first of these (E. Sussex, W. Sussex, Oxfordshire, Herefordshire and Worcestershire, Suffolk and N. Yorkshire), there exists a strong private/state divide but no development of a continuum, whereas the other (Staffordshire, Nottinghamshire, Humberside, Cleveland, Durham and Northumberland) contains very little private representation at all, so there is hardly a divide, let alone a continuum.

There are obviously many LEAs between these extremes, but this classification of extreme types has served to show the variety of local circumstances. This means that for those parents who can make choices about where their children will be educated, the range of choices which they face differs greatly according to where they live. With regard to school rolls, it means that some LEAs will have a much more difficult time planning their future provision than those, such as the last set of counties, where the LEA schools educate nearly everyone. Such unified authorities can also more easily develop and implement provision for those with special educational needs, pupils whom GM schools do not necessarily welcome.

The finances from central government provide the attraction at present, so much so that some private boarding schools are trying to 'opt in'. They have been losing popularity for some time and, with boarding costs rising faster than inflation and demand dropping because of the early 1990s' recession, they too have been threatened by closure. It will be somewhat ironic if the Secretary of State allows them GMS, with threatened closure from market forces getting added to threatened closure from planned efficiency gains as the main stimuli for schools to request the status. It would suggest that any LEA schools threatened with closure by loss of popularity and thus of resources should try the GM route. Such applications by private boarding schools suggest the degree of flux in the present system. Until it settles down, it will be almost impossible to overview provision and rationalise according to demographic needs.

Conclusion

Demographic change is clearly a major input into the educational system. Its effects are often mediated by a number of other processes that combine in varying ways in different localities. When a very variable set of responses to these mediated demographic changes are superimposed, there is a great diversity of outcomes.

The 1980s saw major changes in educational provision and policy, which involved a completely different overall response to demographic change. From rationalisation and efficiency, there was a change to choice and diversity. The introduction of market forces into the educational system, the atomisation of provision, the redistribution of power to individual schools and central government and the blurring of the private/state divide have increased the amount of uncertainty in the system. The extent of this process, however, does vary from place to place. This uncertainty is illustrated well by the difficulties of predicting applications and enrolment for individual schools and of planning any overall strategy towards the changing demography of an area. The growing spatial diversity in educational provision also makes any nationally developed policy extremely difficult to apply. This alone casts great doubt on the ability of the central body, suggested by the 1992 White Paper on Education, to manage the rationalisation of 'surplus' places.

6. FALL-OUT FROM THE DEMOGRAPHIC TIME-BOMB:

A spatial perspective on the labour force effects of the 'baby bust'

Anne Green and David Owen

The changes in the size of individual age groups that cause the problems for planning school rolls outlined in the previous chapter affect to a greater or lesser extent all areas of public- and private-sector activity which are oriented to a relatively narrow age group of the population. On the surface, it might be considered that the labour force would not be too seriously affected because it draws from all ages above the minimum school-leaving age of 16 years old, which even if excluding those of pensionable age (currently 65 for men and 60 for women in Britain) gives a very broad span. Yet this is far from the case; demographically induced trends in labour supply can be a source of great concern, with fears of higher youth unemployment during periods with higher levels of school leaving and the threat of recruitment problems and wage inflation during downswings in the labour market entry of young people.

This chapter deals with the effects of the downswing in numbers of school leavers resulting from the passage of the post-1965 baby bust through late teenage years and young adulthood. This process was accorded great prominence during the period of rapid national economic growth in the latter half of the 1980s. In 1988 the National Economic Development Office and the Training Commission published a report highlighting demographic trends which would result in a shortfall of young people entering the labour market by the mid 1990s (NEDO/TC, 1988). A subsequent report entitled *Defusing the Demographic Time-bomb* outlined the ways in which employers were responding to these changes (NEDO/TC, 1989). Although the salience of this issue has weakened during the early 1990s recession, it is likely to re-emerge with the next economic recovery because of the seemingly permanent move towards lower birth rates and because of locational mismatch between labour demand and supply.

This chapter is concerned primarily with the spatial dimension of this demographic time-bomb and its implications for recruitment strategies, labour supply composition and patterns of labour surplus and shortfall around the country. It begins with an outline of the baby bust and its likely impact on the national labour force and goes on to examine its main features at regional and local scales. Then follows a review of the consequences of these demographic changes for labour supply, with special emphasis placed on changes and variations in workforce composition. Implications of labour supply changes for selected sub-groups are outlined: namely the unemployed, young people, the ethnic minorities, older workers and women.

The baby bust and its implications for the national labour force

The post-1965 baby bust began to impact on Britain's labour market in the early 1980s. It ushered in a 12-year period of continuous and rapid decline in numbers reaching the end of their compulsory schooling and gave rise to the prospect of a major reduction in the size of the younger working-age groups. Whereas the population aged 20–34 years grew by approximately 2 million between 1970 and 1990, over the following twenty years to 2010 it is expected to contract by 2.4 million. In fact, because the number of births rose somewhat after reaching a low point in 1977, the number of new entrants to the working-age population stops falling by the mid 1990s, but this birth rate recovery was relatively modest and restores only about one-third of the earlier contraction.

Change in the size and age structure of the working-age population (the so-called 'population effect') is, however, not the sole determinant of the changing size of the labour force, though obviously it is a key element. The term 'labour force' refers to those people who are in employment or who are identified as looking for work and available to start work (whether or not they claim benefits as unemployed), so the size of the labour force also depends on the proportion of each age group that is in one or other of these categories, known as the 'economic activity rate' or '(labour force) participation rate'. Changes in this proportion produce what is known as the 'activity rate effect' component of change in labour supply. Activity rates vary considerably in the short term because of fluctuations in the pressure of labour demand and in the longer term because of changes in the overall structure of the labour market, so the activity rate effect is more difficult to forecast than the population effect (Spence, 1990). In general, however, over the past twenty years the national picture has been dominated by trends towards substantial increases in activity rates for women, particularly those aged 25–44 years, and smaller reductions for men, particularly those of older working age.

Statistics on past and anticipated labour force change indicate the significance of the baby bust and the extent to which it is mitigated by changes in activity rates. In the first place, a marked reduction in the rate of labour force growth is expected. Over the seven years 1981–88 the civilian labour force grew by 4.5 per cent, some 0.64 per cent a year on average, but for the period 1988–2000 this drops to 3.7 per cent, or only 0.31 per cent a year – less than half the previous rate. The components of this growth also alter substantially because, while the population effect accounted for virtually all the growth in 1981–88 (4.2 out of the 4.5 per cent overall growth), it is the activity rate effect which is expected to contribute the greater part of the subsequent increase (2.0 out of the 3.7 per cent rise).

These developments are accompanied by major changes in the gender and age composition of the labour force. The growing importance of the activity rate effect guarantees a quickening of the trend towards the increasing feminisation of the labour force, as activity rates are predicted to continue falling for men and rising for women. By 2000, women are expected to make up 46 per cent of the labour force, compared with 43 per cent in 1988 and 40 per cent in 1981.

The overall effect on the age composition of the workforce is particularly dramatic. Figure 6.1 demonstrates the marked reduction forecast for the proportion aged 16–24 between 1987 and 2000 and the substantial growth of the 35–44 and 45–59 age groups. In absolute numbers, increases of 17 and 19 per cent respectively are expected for these two older age groups, whereas the number of 16–24 year olds in the labour force is calculated to contract by almost 22 per cent from its 1987 level. The number of 25–34 year olds in the labour force is anticipated to rise by 6.7 per cent over the 13-

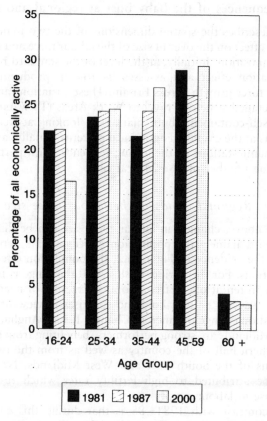

Fig. 6.1   The changing age structure of the economically active population in Great Britain, 1981–2000
(*Source*: calculated from Department of Employment projections of regional economic activity rates (Department of Employment, 1991)).

year period, producing only a marginal increase in its share, but this picture will change significantly during the following decade as the smaller 16–24 year old cohort feeds through into this age band.

From this evidence it can be seen that the significance of the demographic time-bomb lies not so much in its effect on the growth of the overall labour force – though it does produce some slow-down compared with the 1980s trend – but in its effect on the availability of people to take the jobs that employers have been used to placing young people in, notably low-wage jobs and those at the base of any career ladder. The general implications of this type of labour shortage are becoming more widely appreciated and are discussed later in the chapter. The next section is concerned to make the point that some places are going to be hit harder than others by the contraction in the number of young adults and the associated process of labour force ageing. Given that most workers are recruited fairly locally, some employers are going to be affected more than others by such spatial differentiation.

The consequences of the baby bust at regional and local scales

This section describes the spatial dimensions of the two main aspects of the baby bust: the effect on the overall size of the labour force and the implications for its age composition. Initially, at the level of the standard region, it focuses on the population effect and assesses its role in producing variations in overall labour force growth across Britain. These variations are then explored at the more detailed scale of Travel-to-Work Areas (TTWAs) – areas which are relatively self-contained labour markets – looking at trends for types of areas as well as at the overall geographical patterns. Finally, attention is given to the local manifestations of the ageing tendency, with particular emphasis on the changing number of 16–24 year olds.

Regional trends in the labour force size

As mentioned above, changes in labour force size can be disaggregated into two components, namely the population effect and the activity rate effect. According to the evidence of Figure 6.2, the population effect exhibits clear regional variations. For the period 1988–2000 the range is from an increase of around one in ten at one extreme to a contraction of nearly 5 per cent at the other. The general pattern holds no surprises in view of the patterns of overall population change outlined in Chapter 1. East Anglia, the South West and the East Midlands are strong performers, benefiting from net in-migration from the northern half of the country as well as from the two conurbation-centred regions of the South East and West Midlands. Northern Ireland's growth can be attributed to high fertility rates which result in a strong natural increase in labour supply.

The main contrast with 1981–88 is that during this earlier period the demographically induced changes in the labour force were positive for all regions, whereas subsequently five of the eleven regions are expected to record a net decline in labour supply as a result of the population effect. The North West, Northern Region and Scotland appear to suffer the biggest cutbacks

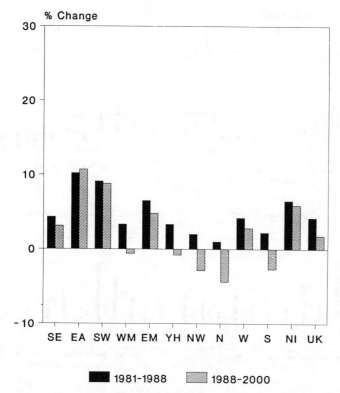

% Change

1981-1988   1988-2000

Fig. 6.2 The population effect on labour force change, 1981–2000, by region (*Source*: calculated from Department of Employment projections of regional economic activity rates (Department of Employment, 1991)).

between the two periods; the West Midlands and Yorkshire & Humberside also record decreases. East Anglia and the South West appear least affected, mainly because of a relatively strong natural increase resulting from previous waves of in-migration by young workers and their families. Even for these two regions, however, the annual rate of growth for the 12 year projection period will be barely half that recorded for the seven years 1981–88.

The activity rate effect is considerably more complex, not least because – unlike with the population effect – it can be understood only by separate treatment of the very different patterns shown by men and women. As Figure 6.3 shows, virtually all regions share the national pattern of declining male rates and rising female rates in both periods, the only exception being a very small increase in the South East's male rate for 1988–2000. At the same time, however, a broad distinction can be drawn between the more prosperous regions and the lagging ones, with the former experiencing faster increases in female rates and the slower decreases in male rates. The contrast between the South West and East Anglia, on the one hand, and the northern regions, on the other, is particularly marked for 1981–88, though less clear for the subsequent period, especially for women.

Given that the activity rate effect tends to reinforce the population effect at

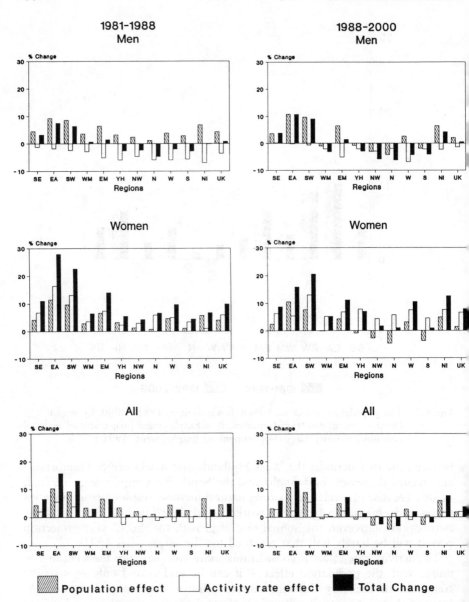

Fig. 6.3   Components of change in civilian labour force, 1981–2000, by region
(*Source*: calculated from Department of Employment projections of regional
economic activity rates (Department of Employment, 1991)).

this regional scale, their combined effect on overall rates of labour force
growth (shown by the solid black bars in the bottom panel of Figure 6.3)
again emphasises the distinction between north and south. As in 1981–88,
so for 1988–2000 the strongest rate of labour supply growth is expected to
be recorded by East Anglia and the South West, followed by the East

Midlands and the South East, plus Northern Ireland which features strongly because of its population effect. At the other extreme, the Northern Region, North West and Scotland are projected to see their workforce shrink; there the growth in female activity rates is not sufficient to offset the decline in male activity rates and the strong negative population effects.

Local changes in labour force size

The range of variation in rates of labour force growth is even wider at the local scale, with significant differences between different types of localities as well as between North and South. Table 6.1 uses a 14-fold spatial and hier-archical classification of TTWAs devised by Green and Owen (1990a). Between 1981 and 1987 all seven categories of TTWAs in the South experienced growth, with the labour force expanding by more than 12 per cent in the Towns and Rural Areas. Rates of growth declined with increasing size of TTWA, but even in London the labour force grew by 3 per cent. In contrast, the labour force of the Northern Million Cities declined, and the rate of growth in the smaller TTWAs was far slower in the North than in the South. Growth rates were highest in the Small Towns and Rural Areas, where the labour force grew at more than twice the rate of the larger TTWAs.

The overall rates of labour force growth projected for the period 1987–2000 (Table 6.1) represent a slight moderation in the annual average rates for most Southern TTWAs. The labour force is projected to grow most rapidly in the Medium-sized Cities there, closely followed by the smaller TTWA classes, and least rapidly in London, though the latter shows an actual increase in annual rate compared to 1981–87. In the North, the labour force is projected to decline in the Million Cities and Subdominant Cities. This represents a continuation of the experience of the 1980s for the former, but a reversal of

Table 6.1   Change in labour supply for Travel-To-Work Area types, 1981–2000 (per cent for period)

TTWA type	1981–87		1987–2000	
	South	North	South	North
Million Cities	3.0	−0.9	6.7	−4.0
Large Dominants	7.9	0.8	13.5	1.4
Subdominant Cities	6.1	1.3	9.3	−3.3
Medium-sized Cities	11.3	2.0	16.9	2.5
Subdominant Towns	13.2	1.6	13.8	1.3
Small Towns	12.1	3.4	16.4	6.1
Rural Areas	12.3	5.0	15.8	6.4

Note: Data relate to percentage change in numbers of economically active persons. The rates for Great Britain are 4.1 per cent for 1981–87 and 5.7 per cent for 1987–2000. 'South' refers to the South East, East Anglia, the South West, and also the southern portions of the East and West Midlands, 'North' to the remainder of Great Britain. For further details of TTWA types, see Green and Owen, 1990a.
Source: calculated from Department of Employment (1991) regional economic activity rate projections and OPCS/GRO(S)/Welsh Office 1987-based population projections.

trend for the latter. The inverse relationship between rates of growth and urban status is projected to be largely maintained, and labour supply again increases most rapidly in the Rural Areas.

Change in labour supply by age group

As noted earlier, the most significant feature of labour force change nationally is the decline in the numbers of 16–24 year olds. Given the very different rates of overall labour force change round the country, particularly through the population effect, it is not surprising to find great differences between places in the actual rate of change in this young group and in its effect on the size and ageing of local workforces. The estimated percentage change in the numbers of economically active 16–24 year olds over the period 1987–2000 is mapped at the TTWA scale in Figure 6.4, which thus presents the spatial manifestation of the demographic time-bomb. It can be seen that in some areas the number of 16–24 year olds in the labour force in the year 2000 will be less than two-thirds of its 1987 level, while in others an increase of up to 16 per cent is predicted.

As in previous analyses, the map seems to reveal both North–South and urban–rural dimensions to the pattern of change. Relatively few TTWAs located south of a line between the Severn estuary and Lincolnshire experience the highest rates of decline, Dorset being perhaps the most concentrated area of labour supply contraction. Most TTWAs along the south coast, in East Anglia and in the south west peninsula fall in the higher quintiles, experiencing the least severe rates of decline or some increase in the number of young workers. These areas contain some of the few TTWAs with projected increases: 16 per cent in Minehead, 4 per cent in Lowestoft and over 2 per cent in Wisbech. London falls just above the median class, with areas of faster decline to the north and east, and areas with slower rates of decline to the west and south. The area of rapid population growth on the South East/East Midlands border from Peterborough to Milton Keynes appears as an island of slower decline in the number of economically active 16–24 year olds. North of the Severn-Lincolnshire dividing line, rates of decline tend to be higher. Scotland appears to be most severely affected by the contraction in young workers, many TTWAs experiencing rates of decline of 40 per cent or more. Almost all of southern Scotland falls into the category of fastest decline, as does most of the West Midlands. The lowest rates of decline are found in the more peripheral areas, notably parts of north-west and south-west Wales, the Welsh Marches and northern and western Scotland.

The main geographical dimensions of these changes are summarised in the first column of Table 6.2, using the 14-fold classification of TTWAs. This analysis emphasises the greater impact of the demographic time-bomb upon local labour markets in northern Britain. Between 1987 and the year 2000 the number of economically active 16–24 year olds is projected to decline by nearly one-third in the Northern Million Cities, and by over one-quarter in the North's Large Dominants and Subdominant Cities. The rate of loss declines with size of TTWA, but even the Rural Areas and Small Towns in the North face a reduction of more than one-fifth in their numbers of

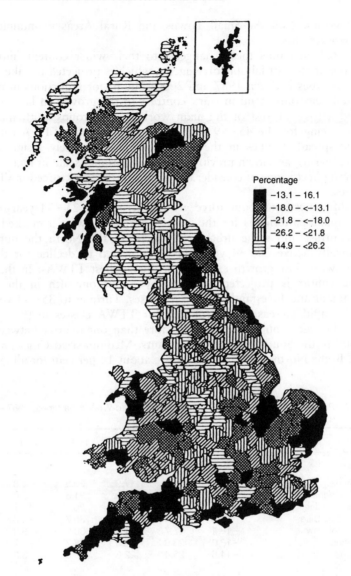

Fig. 6.4   Change in the economically active aged 16–24, 1987–2000, by Local
Labour Market Area
(*Source*: calculated from OPCS/GRO(S)/Welsh Office 1987–based population projections and Department of Employment regional economic activity
rate projections).

economically active 16–24 year olds. In the South, rates of loss are less
extreme, with the highest rates of loss found in the larger cities and the
Subdominant TTWAs within the metropolitan regions; however, even these
TTWA classes experience a slower percentage loss of young workers than
Northern Rural Areas. Rates of decline are again least in the smallest urban

size categories, with the Small Towns and Rural Areas maintaining their young workers best.

Table 6.2 also puts these changes into their wider context, illustrating clearly the process of labour force ageing which is projected to take place in all TTWA types between 1987 and 2000. The major reductions in numbers of 16–24 year olds stand in stark contrast to the growth in labour supply expected across the rest of the main working-age groups, with the largest increases being for the 45–59 year old group for most TTWA categories. The widespread declines in the 60+ group are relatively insignificant in numerical terms, as shown previously for the national scale in Figure 6.1, so the overall effect is one of considerable ageing of the workforce for all TTWA categories.

The differences in labour force growth across the 25–34 year-old span reveal similar patterns as for the 16–24 year olds, with a marked North-South contrast and some urban-size effects. In the North the number of economically active 25–34 year olds is projected to decline for the large TTWAs, with slow growth expected in the smallest TTWAs. In the South, this age cohort is projected to grow by about one-fifth in the smaller Subdominant and Freestanding TTWA classes. Turning to 35–44 year olds, relatively rapid increases are projected for TTWA classes in the whole of Britain; this age group increasing by more than one-quarter between 1987 and 2000 in the Southern Large Dominants, Medium-sized Cities and Small Towns. In the North, rates of increase are about 14 per cent for all classes of

Table 6.2 Projected change in labour supply for Travel-to-Work Area types, 1987–2000, by age group (per cent)

TTWA type	16–24	25–34	35–44	45–59	60+	All ages
South						
Million Cities	−17.4	4.1	17.9	20.3	−2.8	6.7
Large Dominants	−19.0	15.6	25.9	31.5	2.6	13.5
Subdominant Cities	−19.7	5.5	18.8	26.8	13.7	9.3
Medium-sized Cities	−15.6	23.4	28.0	30.7	−1.6	16.9
Subdominant Towns	−18.4	17.7	21.8	30.9	8.9	13.8
Small Towns	−13.9	23.8	26.0	28.7	−1.7	16.4
Rural Areas	−14.6	25.4	22.8	29.0	−3.0	15.8
North						
Million Cities	−31.4	−5.1	13.8	7.2	−11.4	−4.0
Large Dominants	−25.3	1.7	14.6	14.2	−14.3	1.4
Subdominant Cities	−26.5	−4.1	8.3	7.4	−12.4	−3.3
Medium-sized Cities	−22.9	3.9	13.7	13.4	−10.3	2.5
Subdominant Towns	−23.0	4.6	8.7	13.3	−5.6	1.3
Small Towns	−21.8	13.5	15.9	14.9	−8.5	6.1
Rural Areas	−22.1	12.5	17.2	16.7	−3.8	6.5
Great Britain	−21.8	6.7	17.0	19.0	−3.6	5.7

Note: Data relate to percentage change in numbers of economically active persons, disaggregated by age group. See Table 6.1 for details of TTWA typology.
Source: as for Table 6.1.

TTWA except the Subdominant Cities and Towns, in which this age group is expected to increase about half as fast.

A greater contrast is projected for the 45–59 year old element of the labour force. In the South, this element is expected to grow by about 30 per cent in all types of TTWA except the Million Cities, in which the increase is just over one-fifth of the 1987 total. In the North, this section of the labour force is expected to expand about half as fast as in the South, and most slowly in the Million Cities and Subdominant Cities.

In sum, therefore, Table 6.2 provides a powerful insight into the variety of experience expected as a result of the demographic time-bomb working its way into the labour supply of different places around the country. First, there are a number of places where it would appear that the labour force in the year 2000 will be measurably smaller than in 1987, these particularly being the North's Million Cities and other large cities within their area of metropolitan influence. Secondly, several TTWA types look like experiencing slow overall growth in labour supply, up to the national average figure of around 6 per cent; namely all the remaining Northern categories plus London. Finally, all the Southern TTWA types besides London are projected to see significantly above-average growth. Yet even in this latter grouping of TTWA types – and even more so for the other two – there is going to be a very marked reduction in 16–24 year olds, which alongside the general growth of the other three main components of labour supply means a substantial ageing of the workforce by 2000. At the same time, it should be noted that the typology used in Table 6.2 obscures a further degree of local variation that may be expected at the level of individual TTWAs, as can be inferred from the pattern of 16–24 change rates for individual TTWAs shown in Figure 6.4.

Implications of labour supply changes

Interactions between labour supply and labour demand

What is crucial for prospects of employment and unemployment at the local scale is the way in which changes in labour supply and labour demand interact. An increase in labour supply may not result in an increase in the level of unemployment if employment is also increasing. By the same token, a downturn in labour supply may not necessarily result in lower levels of unemployment if the employment base is contracting at a faster rate. Moreover, these straightforward quantitative interactions are likely to be complicated by qualitative mismatches in the characteristics of the labour available and the skills needed to undertake the available jobs. Both increasing and decreasing levels of unemployment may be associated with job growth/loss. Indeed, as indicated above, changing employment prospects may also induce important labour supply responses; such as in-migration and participation increase in the case of employment growth, and out-migration and withdrawal from the labour force in a situation of localised job loss.

The separate effects of demographic, participation, employment and migration change upon the imbalance between changes in labour supply and demand between two dates, as captured by the job shortfall/surplus, may be estimated using the technique of labour market accounting (Green and Owen, 1991). Analysis of aggregate patterns of job shortfalls and surpluses at the TTWA scale reveals marked local variations in the extent of shortfalls/ surpluses. Despite intra-regional variations in experience, a north–south divide is apparent overall, with larger shortfalls/smaller surpluses recorded in northern TTWAs than in comparable TTWAs in the South. Moreover, it is possible to classify areas according to different patterns of inter-relationship between natural change, participation rate change, employment and unemployment change and net migration. In TTWAs such as Liverpool, Glasgow, Sunderland and some coalfield areas, characterised by the largest job shortfalls in the 1980s, there was net out-migration, unemployment increase and either with-drawal from the labour force or smaller than average participation rate increases, in the face of severe employment decrease. Other TTWAs where demographic pressure played a particularly important role in fuelling labour supply increase are identifiable: many of the new towns around London with relatively youthful population structures fall into this category, along with TTWAs from the East Midlands (such as Leicester) and West Yorkshire (such as Bradford) with concentrations of ethnic minority populations (Green and Owen, 1991).

The labour market accounts components may be useful in identifying and assessing the potential role of possible solutions to labour supply shortages. Is it to be expected that unemployment will fall or participation rates increase as a response to labour shortages? Is there a role for internal migration in alleviating the effects of the local-level demographic time-bomb? Or will employers relocate to less pressurised labour markets? Conversely, having considered local demand trends, are economic prospects so bleak that very few areas will have labour shortages? Some of these issues are addressed in the remainder of this chapter, alongside consideration of the implications of labour supply changes for selected workforce sub-groups.

The unemployed and the role of migration and commuting

In a context of labour shortages and a downturn in the numbers of young people, there would appear to be considerable scope for re-employing the unemployed. Indeed, a downturn in unemployment did accompany an expansion in demand at the national scale in the second part of the 1980s, but nevertheless unemployment remained at a high level and is forecast to increase again over the short-term to 3 million (IER, 1991).

At the local level there are considerable differentials in the incidence of unemployment and long-term unemployment (Green and Owen, 1990b), with inner city areas and the large metropolitan areas of northern Britain and Northern Ireland being particularly hard hit. The current recession has seen some of the fastest increases in unemployment in localities in southern England, but nevertheless unemployment rates in these areas remain lower than in many other parts of the UK, while some of the highest unemployment rates are still to be found in the traditionally depressed areas of north-east England, west central Scotland, the south Wales valleys, and Merseyside, along with parts of South Yorkshire and the manufacturing heartland of the West Midlands.

In the light of such variations in local unemployment rates, is there a role for labour migration in reducing imbalances? Evidence suggests that in the absence of labour migration local unemployment differentials would have been even more marked. Nevertheless, those unemployment differentials remaining represent only to a limited extent the outcome of spatial mismatches in labour demand and supply. Housing market rigidities — in the form of inter-regional house price differentials, a depressed owner-occupied sector, a reduction in the size of the private rented sector, and barriers within the public rented sector do act as a brake on long-distance labour migration (Green et al., 1986, Allen and Hamnett, 1991), reducing the scope for alleviating such spatial mismatches.

If long-distance migration is difficult, is there scope for encouraging commuting or altering patterns of commuting to alleviate local labour supply and demand imbalances? Evidence suggests that many of the unemployed could benefit from broadening their spatial horizons (e.g. Quinn, 1986). Interestingly, from an opposing perspective, Training and Enterprise Councils in South East England have begun exploring some of the options for addressing local skills gaps by encouraging commuters to work locally rather than commute to London (IFF, 1992).

There are two broad views on the relative significance of structural mismatches in generating local unemployment differentials. On the one hand, there is evidence to show that the unemployed are disproportionately drawn from the less skilled occupational groups in declining industries (White, 1991). On the other, surveys have shown that there is a wide range of vacancies that could be filled by many of those who are unemployed. A number of local surveys (e.g. Meadows et al., 1988) have revealed that high levels of vacancies coexisting with a high level of unemployment cannot be explained solely by the types of vacancies available being unsuitable for the unemployed, indicating that in many cases there is scope for alleviating local labour shortages by turning to the unemployed.

Young people

The fall in the number of young people entering the labour market has meant greater competition amongst employers for those who are available, although many employers remain confident of their ability to 'compete' for a dwindling number of well-qualified young people — even in tight labour markets (Waite and Pike, 1989). Attempts to increase staying-on rates in education and training are likely to further reduce the number of young people in the workforce. As the downturn in numbers of school leavers available for work became apparent in the late 1980s at a time of national economic recovery, skills shortages emerged on the policy agenda, and there were fears that these skills problems would intensify as a result of the demographic time-bomb. The 1988 White Paper on 'Employment for the 1990s' (Department of Employment) made much of the downturn of young workers and the need to look beyond traditional recruitment pools. Indeed, there is no doubt that some industries with a traditional reliance on young workers, such as nursing, faced particular difficulties. However, some believe that the role of the demographic time-bomb in contributing to these skills shortages has been exaggerated, instead suggesting that more attention needs to be paid to patterns and processes of labour market discrimination. According to Haughton (1990, p. 343): 'The demographic time-bomb seems much less important to skills mismatch than the rhetoric suggests. It is a temporary problem whose scale and extent have been commonly exaggerated'.

There is no doubt that there has been geographical, industrial and occupational unevenness in moves to recruit new groups, in the light of a reduction in young people entering the labour market and skills shortages – both of which have varied in severity between localities. In general, at the height of the economic boom, recruitment difficulties were most severe in London, the South East, the South West, and other southern regions, and in response employers there adopted a wider range of options, such as increasing pay, accepting part-timers, widening recruitment methods and retraining (Smith, 1990; Green and Hasluck, 1991). At a time of recession these more innovative employers may abandon some of these measures and adopt behaviour akin to employers in those areas where fewest concessions or changes were made in the period of most intense labour supply problems nationally.

Ethnic minorities

The non-white population is characterised by a much younger age structure than the white population of Britain. In 1989 nearly one-third of the non-white population were aged under 16 years, compared with under one-fifth of the white population. The non-white population is also relatively more concentrated in the 16–24 and 25–34 age groups. By contrast, only 8 per cent of non-whites are aged over 60 years, compared with over one-fifth of the white population. Indeed, it has been estimated that approximately half of the increase in the UK population over recent years has been accounted for by expansion of these groups, though they account for no more than 5 per cent of the total population (Haskey, 1991). An increasing proportion of young entrants to the labour market are from such groups, and ethnic minorities will account for an increasingly large proportion of the labour force over the next decade.

The black population displays an uneven spatial distribution (Haskey, 1991), being concentrated in some of the larger cities, such as London and Birmingham, and the older manufacturing towns of northern Britain. Some of these areas have witnessed large employment declines over the 1980s, and unemployment rates among some ethnic minority groups are higher than those of their white counterparts, while their economic activity rates are lower (Owen and Green, 1992): in 1985–1987 it was estimated that the average economic activity rate for ethnic minority groups was 66 per cent, compared with 79 per cent for whites. At the intra-urban scale there are even more pronounced concentrations of people from the ethnic minority groups: in London, for example, the ethnic minority share of the total population was estimated to exceed one-fifth in Hackney, Haringey, Kensington & Chelsea,

Lambeth, Newham, Tower Hamlets, Wandsworth, Westminster, Brent and Ealing in the late 1980s, while in Barking & Dagenham, Bexley, Bromley, Havering and Richmond-upon-Thames the proportion was less than one-tenth (Haskey, 1991). Similarly, in Coventry a large-scale skills audit showed the non-white population to be concentrated overwhelmingly in two inner city wards (Elias and Owen, 1989).

In many localities characterised by a greater than average representation of residents from the ethnic minorities, the downturn in the number of young people has been less severe than elsewhere. Moreover, in such areas the ethnic minorities account for an increasing share of young labour force entrants; for example, in Bradford it is estimated that the proportion of school leavers of Asian origin will rise from 24 per cent in 1990 to 31 per cent in 1993 (Employment Department Yorkshire and Humberside, 1991).

It is clear that despite individual successes, black people have become trapped at the bottom end of the hierarchy of jobs, income and status. Black immigrants initially came to the UK as a replacement labour force, occupying the least skilled, the dirtiest and the lowest paid jobs. Even if lower formal educational qualifications of black people than white people are taken into account, they still do less well in the labour market than might be expected. Persistent inequalities in employment, over and above those expected on the bases of education and previous job-training, can mainly be accounted for by discrimination (Hudson and Williams, 1989). It appears that employers are not drawing on the full potential of ethnic minority groups.

Older workers

As older workers proportionately account for a larger part of the labour force, the question of employing them has become an important one for many employers. In the past there has been a particular emphasis on training the young (Training Agency, 1990); it is clear that on the basis of projected demographic, technological and economic trends, in the future more training will be required for adults already in the labour market.

Just as activity rates amongst the young are declining, so are participation rates at the other end of the age range. The proportion of people continuing to work beyond the state pension age has been falling for many years. This appears to have been associated with the rise in the value of state pensions, in the value and coverage of occupational pensions and the value of capital assets in people's possession. In recent years employment opportunities for older workers below the state pension age have also been declining, as early retirement has been used as a popular method of achieving workforce reduction. The economic activity rate of males aged 60–64 years declined from 91 per cent in 1961 to 82 per cent in 1975 and 55 per cent in 1990. Figures quoted by the CBI (1988) appear to indicate that the majority of the over 50s currently not in work are either too ill to work or permanently retired, but research suggests that a substantial proportion of the overall increase in disability and decline in economic activity among older men is in fact attributable to the general rise in unemployment.

Case study research indicates that in the face of a decline in the number of young workers, recruitment difficulties and high labour turnover, hiring of older workers may prove a viable option in particular local contexts. Evidence from the DIY chain B&Q, and in particular its Macclesfield store which was staffed entirely by employees aged over 50 years, revealed that in comparison to their younger peers, older workers were absent less often, just as productive, equally flexible, no more costly in wage terms and just as willing to train (Hogarth and Barth, 1991). However, a survey of over 300 large private and public sector organisations conducted in autumn 1991 showed that age is still considered an indicator of suitability, with some employers anticipating problems with older workers and few targeting older workers to meet labour shortages (Employment Department Group, 1992).

Women

In the context of a decline in the numbers of young people and economic expansion at the national scale in the late 1980s, particular attention has been, and will continue to be, focused on women as an under-utilised resource (Employment Department Group, 1991). There appears to be considerable scope for an increase in female participation rates to alleviate local supply shortages, as and when they arise.

The increase in the number of women in the labour force is one of the main dimensions of recent change. Two-thirds of labour force growth over the period from 1983 to 1987 was made up of married women, and it is estimated that women will take up approximately 90 per cent of the new jobs created over the next ten years (Ermisch, 1990). A tendency is emerging for women leaving the labour force for child-birth and child-rearing to regard their break as temporary, and to return to work sooner after child-birth and often between the births of children.

Although there has been an increase in female participation rates in all regions of the UK and virtually in all localities, there are some significant spatial variations in labour force participation. In the 1980s female partici- pation rates were higher in the South East than elsewhere, and this regional differential became even more marked over the decade as the female partici- pation rate in the South East rose to 80 per cent: the South East being the only region to display a female participation rate higher than the UK average. The higher proportion of two-income households in this region helped to sustain and fuel large increases in house prices. Female participation rates have traditionally been higher in urban than in rural areas, with women in the remoter areas often having very limited employment opportunities, compounded by problems of access to the jobs available. Information and communications technologies may have the potential to extend the oppor- tunities available, by enabling activities to be undertaken in a wider range of locations. Nevertheless, women remain concentrated in the traditional service sector industries, and in lower level occupations.

Women returning to work often face considerable problems in balancing their domestic and work roles, and arranging and paying for suitable childcare can be very difficult (Metcalf and Leighton, 1989). Surveys in contrasting local labour markets in the UK indicate that these problems are often compounded by difficulties of access — to work and training (Hardill and Green, 1991; Healy and Kraithman, 1989). In an attempt to attract and retain women employees, some employers have enhanced maternity leave arrangements, introduced career break schemes, expanded training opportunities for women returners, provided assistance with childcare — including childcare vouchers and workplace nurseries, and provided part-time and flexible working hours. Unsurprisingly, employers in southern Britain, facing the greatest supply-side shortages, tended to be at the forefront of such initiatives in the late 1980s. Overall, however, there has been more rhetoric in these directions than examples of good practice, and it remains to be seen whether some of the gains made by women in the late 1980s will be lost until the next economic recovery; in the context of recession economic prospects may seem so bleak that very few localities experience labour shortage.

Conclusion

Not all parts of Britain are experiencing the effects of the demographic time-bomb to the same extent: in general, northern Britain faces a much greater downturn in 16–24 year olds than southern Britain. The reduction also tends to be greater at the upper end of the urban hierarchy. The net effect is that the largest cities in the north and midlands (such as Glasgow and Birmingham) will undergo some of the greatest proportionate declines in the number of 16–24 year olds in the labour force. However, projected increases in the 35–59 year old age groups largely offset the losses of young entrants to the labour force. Despite reductions in numbers of young workers, local labour market areas in southern Britain are projected to experience increases in the economically active population in excess of 10 per cent by the year 2000. Indeed, projected growth rates for the 35–59 year old age group are relatively more significant than the decline in the 16–24 year age group. This suggests that the main impact of the demographic time-bomb is on the composition, rather than the size of the labour force.

What is particularly crucial, however, is the way in which changes in labour supply interact with changes in labour demand. In some areas decreases in young people may be more than compensated for by increases in economic activity rates, so obviating labour shortages. In other areas decline in employment may outstrip decrease in the numbers of young people, possibly resulting in an increase in unemployment and/or out-migration. A wide range of permutations of labour supply and demand interactions are possible at the local scale.

The availability of older workers suggests a ready substitute for the declining number of young people leaving full-time education, but the utilisation of this labour resource is dependent upon the adoption of new recruitment practices and training by employers. However, available evidence points to spatial variations in the innovativeness of employers in terms of human resource development. Clearly, many of these older workers will be women, and rising female activity rates in the 25–34 and 35–44 year age groups will be accounted for by women returning to the labour force.

Along with the growing ethnic minority population, another possible source of substitute labour is immigrants, moving freely within the European Community after 1992. To some extent, such in-migration has already taken place into tourist-dominated and other local labour market areas in southern Britain. In-migration of better educated young Europeans might block re-employment of older people, the unemployed, women returners and ethnic minorities in some occupations and industries in some localities.

Since changes in national and local economic circumstances impact upon the labour force in multifarious ways there is a need for continuous monitoring of trends by central government, local government and other local actors, such as chambers of commerce, Training and Enterprise Councils (TECs) and colleges. Obviously, there is scope for updating the types of analysis reported in this chapter when more recent data become available (see Healey, 1991, for a review of possible sources). The 1991 Census of Population is particularly useful in providing information on economic status at the local scale. Information on economic activity and the changing age, gender, industrial and occupational structure of employment is also available on a more frequent basis from the Labour Force Survey, although the spatial disaggregation is more limited than for the Census of Population. A range of demographic sources – including Mid-Year Estimates (at the Local Authority District scale) and population projections (at the county scale) may be accessed via the National Online Manpower Information System (NOMIS). Labour market assessments produced by TECs and LECs provide up-to-date information on the labour supply situation at the local scale, as well as outlining changes in labour demand. Another useful source is the Department of Employment's Employment Gazette; notably the periodic articles entitled 'Labour Force Outlook'. Various briefing papers distributed by the Employment Department Group's Skills and Enterprise Network provide information on ongoing research on labour market issues, including the implications of demographic change.

7. DEMOGRAPHY AND HOUSE-BUILDING NEEDS:

A critique of the 'demographic bulldozer' scenario

Dave King

Since shelter is recognised as one of the most basic of human needs, it might be thought that in an essentially welfare society like Britain there would be a close relationship between housebuilding and population growth. There are several good reasons why this is not the case, including the demand for second homes, the replacement of unfit stock, fluctuations in the level of vacancies and changes in the economic and public-policy contexts which influence building rates. The demographic factor itself is, nevertheless, extremely important, not least because it tends to operate in more complex and subtle ways than is often appreciated.

This chapter provides an insight into this complexity by investigating and challenging the widely held belief that, because Britain's birth rate peaked in the mid 1960s and then fell sharply over the following ten years, the present decade will see the need for extra houses falling to unprecedented low levels. Relief from 'town cramming' pressures and further urban encroachment into the countryside are not as near at hand as many would have us believe. Among other considerations, it is shown that the demand for owner-occupied housing lags behind the need for rented housing because of the higher average age of purchasers and also that 20–30 year olds leaving home to get married are by no means the only source of new household formation. It demonstrates that the locations where new households are forming and where existing households want to move to are not necessarily those where vacancies exist, nor is the available stock always of the right type in relation to size, tenure and cost.

Contrasting views on housebuilding needs

A fundamental shift in attitudes towards housebuilding needs has occurred since the 1960s. At that time demographers and strategic land-use planners joined forces to argue the case for major housing growth across the country as a whole and throughout the south in particular. This was the decade in which Milton Keynes and Peterborough new towns were conceived. The early 1990s, by contrast, is a period in which a number of demographers and economists have pointed to demographic evidence as heralding a period of substantially reduced housebuilding pressure. Seeing parallels with the 'demographic time bomb' in the labour market described in the previous chapter, the phrase 'demographic bulldozer' has been coined (see, for example, Ermisch, 1990).

The 'demographic bulldozer' idea has fallen on receptive ears because it potentially solves a difficult political conflict inherited from the Thatcher years. On the one hand, the 1980s ideological viewpoint emphasised the promotion of individual choice and the 'free market', which saw the private housebuilding industry as the most effective and efficient means of meeting the interests of the consumer and supported it through changes in the operation of the town and country planning system as well as through large reductions in public-sector housing investment. On the other, successive Conservative governments have become increasingly embarrassed by the electorate's responses to environmental protection matters, most notably the NIMBY (Not In My Back Yard) instincts of the Tory shires. In these circumstances, it is not surprising that politicians and land-use planners alike should take some comfort from suggestions of an incipient reduction in housebuilding pressures.

A contrasting view, however, is posed by many housing professionals (e.g. Bramley, 1989; Wilcox, 1990; Institute of Housing, 1992). They point to ever increasing rates of homelessness, ever growing council house waiting lists, the poor living conditions of very large numbers of households, and the current unmet aspirations of vast numbers of young adults living with parents and wanting to enter the housing market, as demonstrated by the results of Housing Needs surveys (e.g. London Research Centre, 1988). Whereas the 'demographic bulldozer' advocates point to a continued reduction in new housebuilding rates in the future as a result of the demographic effects, to rates below 100,000 per annum nationally, possibly tailing off to zero by 2025 or so (Ermisch, 1990), the housing professionals' viewpoint argues strongly for increases in housebuilding over the next decade. A commonly quoted figure is the need for at least 100,000 new 'affordable' (i.e. social) rented dwellings alone per annum, i.e. excluding 'market' housing, which might be expected to double those numbers, according to the Institute of Housing (1992). When matters of existing living conditions and frustrated aspirations are added to the equation, a new-build figure of nearer 400,000 per annum can be identified for the 1990s (Niner, 1989). We might refer to this as the 'housing needs' scenario.

The geographical implications of the 'pure' version of these viewpoints differ starkly. The 'demographic bulldozer' scenario is believed to offer the prospect of the 'containment' of new development within existing metropolitan and urban frames, allowing a virtual fossilisation of the fabric of rural Britain. Though this policy approach would cut directly across established dynamic patterns of migration, commuting and employment, under this scenario the threat from large-scale housing development would gradually recede and a more flexible and responsive planning environment could prevail. By contrast, the 'housing needs' scenario recognises that Britain's metropolitan and large urban areas have a large enough backlog of housing needs of their own to cope with without being seen as the solution to the housing growth pressures of shire counties as well. Any concerted attempt at meeting this scenario would also have to recognise that housing needs are not simply restricted to more urban localities but are spread across much of non-metropolitan and rural Britain (Bramley and Paice, 1987).

The actual course of future events depends partly on which scenario describes most realistically the level of housebuilding needs and partly on the manner in which the political and planning systems respond to it. The primary purpose of the remainder of this chapter is to throw light on the demographic aspects of these divergent viewpoints. In order to investigate the interplay between demography and housebuilding needs, the components of the 'demographic bulldozer' effect are outlined first, and then the location of growth in household numbers is examined in terms of both 'indigenous' growth and the impact of migration. After this, a more detailed look is taken at the relationships between demography and the housing market, exploring the complexity of translating forecasts of changing household numbers into estimates of housebuilding rates. Finally, the role of the land-use planning system intervening in the locational distribution of households is examined, raising issues not only about the responsiveness of the planning system but also about the commitment of government to assessing housing need and making arrangements to meet it.

The demographic bulldozer

The 'demographic bulldozer' effect refers to the situation where the national growth of households substantially reduces over time. Because households are the basic demographic units which occupy dwellings, it follows that if growth in the number of households falls, all other things being equal, net growth in the number of dwellings can also be expected to fall. Consequently, it is argued, pressure on the housing market will reduce, the stabilisation/ reduction in house prices of the early 1990s might be expected to continue indefinitely and the need for new housebuilding (and greenfield sites for it) will shrink significantly.

Figure 7.1 shows the 'demographic bulldozer' effect using the results of an analysis carried out at the Building Research Establishment for the Department of the Environment (Corner, 1991). It portrays the estimated historical and projected future pattern of net household increases for England and Wales over the period 1961–2011. Up to 1983 the overall pattern is one of steady growth, averaging around 130,000 households per annum, with annual fluc-tuations of 25,000 or so around that mean. During the mid 1980s net household growth increased to around 200,000 yearly, but thereafter it is projected to fall back to a low of around 100,000 per annum just after the turn of the century. It is the slide from the demographic high of the 1980s to a more typical postwar pattern in the 1990s and beyond which has triggered discussion about the demographic bulldozer.

Figure 7.1 also allows the identification of the four demographic components behind this pattern of net increases in households over the period. The first is the size of the adult population which shows a pattern that roughly corresponds

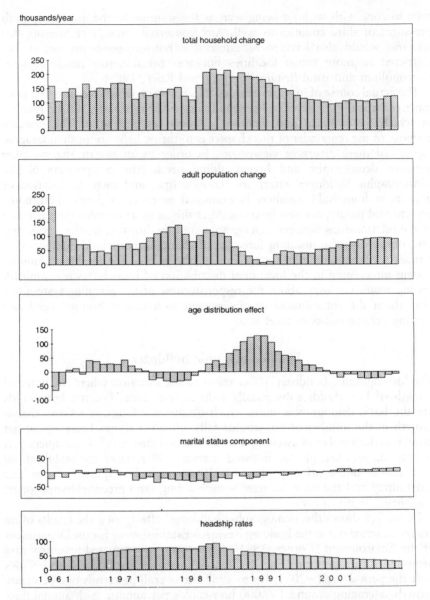

Fig. 7.1 Components of annual change in the number of households in England and
Wales, 1961–2011
(*Source*: Corner, 1991).

to concerns about labour supply shrinkage. The 1960s baby boomers acceler-
ated the growth in the size of the adult population from the mid 1970s, with
the level tailing off during the 1980s. From the 1991/93 low point the
pattern is expected to show steady increases over the next two decades.

The age distribution within the adult population is also a very significant component in numerical terms, running largely counter to the adult population component. As new entrants to adulthoood shrink in number, so the age distribution shifts. The older the population is, the more likely it is to have formed into households, so the effect of the 1970s 'baby bust' cohorts entering the scene from the mid 1980s, combined with the ageing of the 'baby boomers' cohorts, is to lead to a large and substantial peaking of the age distribution effects in the early 1990s. Unlike in the labour market, individuals do not necessarily make the transition into the housing market upon reaching adulthood. The peaking of the baby-boomers' entry into the housing market in the early 1990s, rather than the early 1980s, neatly encapsulates that lag effect.

The last two components relate to the changing propensity to form households. This is influenced by changing marital composition and by changes in headship rates, the latter defined as the proportion of adults (grouped by age, sex and marital status) that act as household heads. The last two panels of Figure 7.1 summarise, respectively, the effects of changing patterns of marriage and divorce, and the steady increase in headship rates over time. Both are clearly less volatile over time than the other two demographic components. The negative effect of the marital composition component relates to an increased tendency for young adults to marry at later ages than in previous decades. In particular, far fewer 20–24 year olds are married now than was historically the case – under 20 per cent in 1991, as compared with 50 per cent in 1971. Interpretation of this component from the 1980s onwards is complicated by the rapid growth in cohabitation for this group (Haskey and Kiernan, 1989). However, the *combined* effect of marital status and headship rate components can securely be taken as a measure of propensity to form households and accounts for an increase of around 50,000 households per annum, both in the past and projected into the future.

This analysis suggests that there is strong evidence of the 1980s and early 1990s being a peak period for net increases in household numbers and that the influence of the changing size and age structure of the adult population has been dominant in that pattern. More significantly in the present context, it also indicates that, although the 1990s show a falling away from that peak, the overall pattern merely returns to the levels of household growth experienced in the 1960s and 1970s rather than involving the catastrophic reduction implied by the phrase 'demographic bulldozer'.

At the same time, it should be stressed that varying degrees of uncertainty exist in all the components examined here. In the medium and long term improvements (or otherwise) in mortality rates are important, yet uncertain, while in the longer term fertility rates are particularly important (though not in the medium term since it is only on reaching adulthood that new cohorts begin to be of interest in this context). Annual fluctuations in the international migration component of the projections can be of the order of plus or minus 50,000 adults, which would translate into perhaps 25,000 households either way. The propensity to form households is the other major factor in the short and medium term but, according to Figure 7.1, constitutes a relatively steady and dynamic element.

The location of growth in household numbers

Given that the increase in household numbers seems likely to continue relatively strongly over at least the next two decades, the central planning issue is the location of this growth and the impact on individual areas around the country. When considered at national level, there might appear to be little ground for controversy; after all, even the higher level of around 200,000 extra households a year in the 1980s represents an annual addition of less than 1 per cent. At subnational levels, however, the growth rarely occurs in proportion to the existing distribution of households; indeed, for reasons concerned with available space as well as other factors, often the strongest growth takes places in relatively less populated areas. At the local level, growth in household numbers is caused not just by the demographic components outlined above but also by migration; and, as we shall see, even the former varies markedly from place to place.

In the context of local household analysis, the term 'indigenous' is used to denote the growth in a locality which stems purely from the local population, assuming no net inward (or outward) migration. In recent years the concept of indigenous growth has become important as a benchmark against which planners and policy-makers can assess whether or not a locality is proposing to cater for the consequences of its own expected growth in household numbers. Yet, over much of the country, it has been migration which has been the dominant factor in the local growth (or decline) in household numbers.

The relative importance of indigenous and migration-induced growth, and the extent to which this varies across the country, is explored in Figure 7.2. For each county in England, this shows (on the vertical axis) the indigenous growth rates for the decade 1991–2001, as projected by the Chelmer Population and Housing Model, and (on the horizontal axis) the growth in household numbers which would result from the continuation of the net migration levels experienced in the 1980s. Together, they produce an estimate of the total change in household numbers for the decade, with the diagonal line in Figure 7.2 representing a 15 per cent rate of overall growth.

It is clear from Figure 7.2 that both components vary widely from place to place. The rate of indigenous growth in household numbers is greatly affected by differences between counties in their inherited age profiles. Counties which have traditionally lost their young adult population through out-migration or gained elderly people through retirement migration show low indigenous growth rates; for example, the Isle of Wight, Devon, Dorset and East and West Sussex. By contrast, many counties which have experienced 'planned' or employment-related growth in the past few decades, drawing in a young mobile workforce, show very large rates of indigenous growth, as the children of these newcomers increasingly reach household-forming age (the so-called 'second-generation effect'); for example, Berkshire, Buckinghamshire, Cambridgeshire and Northamptonshire.

The range of migration-induced growth in household numbers between counties is even wider than for indigenous growth. Under these projections, much of the indigenous growth of London, the metropolitan counties and

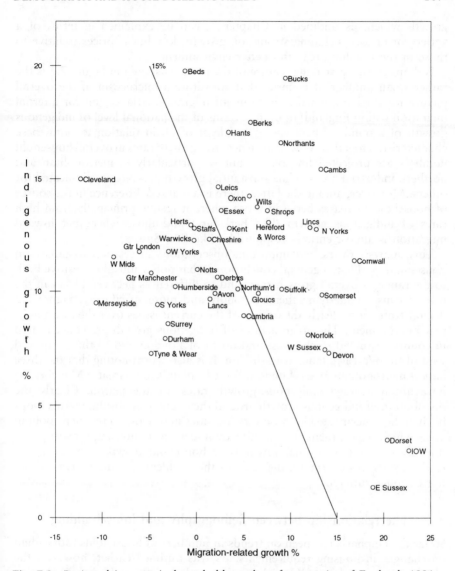

Fig. 7.2 Projected increase in household numbers for counties of England, 1991–2001: indigenous versus migration-related growth
(*Source*: Chelmer Model Projections, Population and Housing Research Group. The diagonal line represents the 15% rate of overall growth in household numbers).

other northern industrial counties can be expected to be decanted to other parts of the country through net out-migration; at the extreme, representing a net decline of over 10 per cent in household numbers for Merseyside and Cleveland over the decade. Meanwhile, some counties can expect migration-induced increases of up to 22 per cent, most notably those along the South Coast like the Isle of Wight, Dorset, East Sussex and Cornwall – a pattern of growth which, as outlined in Chapter 1, can be explained in terms of a variety of factors including strong job growth, low house prices (relative to those in the London area) and better environment.

Perhaps the most striking feature of the patterns shown in Figure 7.2 is the rather small number of counties that constitute a microcosm of the overall picture for England, whether in terms of migration balance (nil for internal migration within England) or even in terms of the national level of indigenous growth of around 10 per cent − and least of all in relation to both these characteristics together. Low or even negative growth rates in overall household numbers are projected for some counties, particularly in metropolitan and northern industrial areas, while substantial growth pressures are indicated for others. Moreover, among the latter, there is a marked difference in the source of household increases between those where it results principally from high rates of indigenous growth (like Berkshire) and those where net inward migration is almost entirely responsible (like East Sussex).

This analysis raises two important observations in relation to policy on housebuilding. First, a general conclusion is that policy must be sensitive both to the range of overall growth rates and to differences between places in the mix of causes underlying them. Secondly and more specifically, Figure 7.2 demonstrates very clearly the origins of the current issues over the location of new development. The greatest rates of indigenous growth are concentrated in counties situated to the west and north of London and to the south and west of the West Midlands conurbation. It is also worth noting that the three largest metropolitan areas (London, West Midlands and Greater Manchester) have about average indigenous growth rates for the period. Clearly the distribution of indigenous growth around the country in absolute terms looks likely to be concentrated in those very localities in or close to the metropolitan areas which regard themselves as already under severe housing growth pressures. Even were the national downturn in household growth from the 1980s to the 1990s to materialise, the scale of the problems for these areas would not be much different.

The relationship between demography and housebuilding

So far the emphasis has been on trends in numbers of households rather than on changes in housing requirements. As was outlined earlier, however, the phrase 'demographic bulldozer' implies a close link with the number of dwellings and, in particular, with the need (or otherwise) for new construction. The purpose of this section is to show that the latter cannot be read off directly from estimates of net change in household numbers and that, on the contrary, the relationship between the two is extremely complex. This can be demonstrated most readily by reference to national data, but the point is even more relevant at local scale because it holds implications not only for the overall volume and timing of housebuilding but also for its geographical location.

The lack of relationship between growth in household numbers and volume of housebuilding is demonstrated in the simplest form in Figure 7.3. It is only since the early 1980s that there has been a reasonably close correspondence;

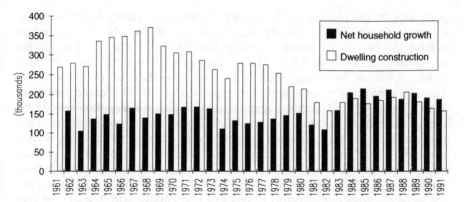

Fig. 7.3 Household growth and housing construction in England and Wales, 1961–91
(*Source*: Corner, 1991; *Local Authority Housing Statistics*, HMSO).

during the 1960s and 1970s construction rates were running much higher not just in relation to the 1980s level but also by comparison with net household growth. The contrast between the two periods can be explained partly by references to the strong public-sector involvement in slum clearance and property replacement in the earlier decades, but this is not the whole explanation. Net household growth has actually been outstripping dwelling construction since the early 1980s, when public-sector completions have been at an all-time low and the private sector has been hard pushed to maintain its average performance of the earlier two decades. Clearly the economics and politics of the housing market and public-sector provision have had a major influence on the level of house completions, independent of demographic effects, and are likely to continue to do so, both in terms of short-term market fluctuations and longer-term policy impacts.

Mismatch problems provide a second example of the complexities involved in translating household change figures into housing requirements. The analysis in Figure 7.3 takes no account of any difficulties in matching households and housing, either by type or geographical distribution. For instance, a surplus of dwellings in Merseyside could not readily be expected to solve a housing crisis in central London, nor can a surplus of bed-sits be expected to solve a local shortage of family accommodation. Mismatch problems are a recognised element of any housing appraisal and may require new construction to resolve them (HSAG, 1977). Such mismatches should be extended to considerations of tenure, house condition and residential environment, as well as size and general location. They must also embrace the concept of affordability and accessibility, since substantial latent demand may exist in any given location. In this context, it should be noted that mismatch problems are not likely to have been getting any less significant in recent years because government strategy of relying on owner occupation has been largely at the expense of the adjustment mechanism historically provided by the rented sector, private and public.

A third aspect relates to tenure split of household formation and, in particular, to the phasing of the impact of cohort-size changes on the owner-occupied sector. It has already been seen that household formation lags behind the process of people reaching adulthood, but the effect on owner occupation comes even later. The propensity of households to own their homes is lowest for young adult heads; only one-third of 15−24 year old heads in 1985 compared to almost three-quarters of 30−44 year old household heads (Table 7.1). The median age of entry into owner occupation was 25 for the 60 per cent of first-time buyers in 1987 who were leaving their parental (or equivalent) home, and even later for those entering the housing market via renting − 29 for the 27 per cent of first-time buyers coming from private-rented accommodation that year and 32 for the 8 per cent moving from public-sector renting (excluding Right to Buy sales). Table 7.2 demonstrates the same phenomenon in terms of types of households, with the 25−34 year olds forming the modal group.

The effect of these patterns of entry into owner occupation is to both lag and smooth out the boom−bust effects on private-sector housebuilding implied by a superficial analysis of demographic change. Whether the new owner-occupying households are moving directly into newly built homes or buying older housing and enabling existing owners to move to them, the consequence is that age-structure-related demand for net additions to the private-sector stock remains at a high level until the mid 1990s − at a net increase of over 80,000 households per annum. This conclusion is in sharp contrast to assertions about the demographic trough for this sector coinciding with the recession of the early 1990s.

Table 7.1 Propensity for owner occupation, by age of head of household, Great Britain, 1985

Age of head of household	% owner-occupiers	Age of head of household	% owner-occupiers
15−24	34	60−64	58
25−29	62	65−69	52
30−44	72	70−79	48
45−59	65	80+	46

Source: General Household Survey 1985 (unpublished tables).

Table 7.2 Age distribution of heads of first-time-buyer households, by type of household, England, 1987 (per cent)

Age of head of household	One Male	One Female	One Male + One Female	Other households
Under 25	32	29	40	34
25−34	48	45	40	46
35 or over	20	27	19	20

Note: columns may not sum to exactly 100 because of rounding.
Source: Building Society Mortgage Survey.

Such an analysis, of course, assumes that the owner-occupation propensities remain unchanged in any given age group. In reality, the past has seen large shifts towards owner occupation in virtually all age groups, and not simply because of council house sales. There is inevitably debate about whether or not the 30–44 year olds have reached a 'saturation' level with regard to affordability issues (Bramley, 1989). Even were this so, however, the experiences of each cohort are now such that generation effects can be expected to continue to drive up the propensities in older age groups, whose current lower levels of owner occupation (see Table 7.1) are principally a reflection of the possibilities and expectations prevailing when they were forming their initial households in the 1950s and 1960s.

There is also the question of trends in headship rates and the way in which these react to changes in the state of the housing market. Figure 7.4 shows the results of applying past recorded changes in age-specific headship rates to the projection of household numbers for the period 1988–2001 and compares them with the Department of the Environment's 1985-based projections. There appears to be general consistency in outcome for the 45–64 age group – not surprising, given that at this point housing career patterns have been largely stabilised, earning power is greatest and therefore accessibility and affordability issues have largely been resolved. For the other age groups, however, the exercise reveals great sensitivity to the alternative assumptions. Arguably this particularly reflects sensitivity to changes in affordability of and accessibility to the rented sectors. This is very evidently true for people under 30, with knock-on effects over time for the 30–44 year old age group. The highest headship rates for 15–29 year olds can be found in metropolitan bed-sit land, for instance, where the housing is relatively accessible. The availability of affordable accommodation of an appropriate type and location

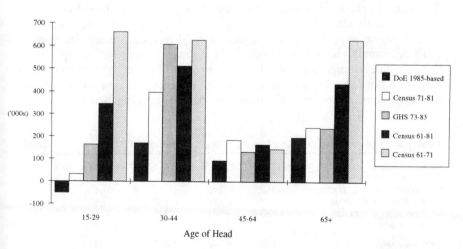

Fig. 7.4 Projected net increases in household numbers in England, 1988–2001, by age of household head: sensitivity to alternative headship rate assumptions (*Source*: King, 1991. DOE = Department of the Environment; GHS = General Household Survey).

may also play a significant part in the sensitivity shown by the over 65 year old group (King, 1991).

Finally, comment must be made about changes in household composition and their implications for the matching of household and dwelling types. The professional literature has made much of the net reduction in married couples and the growth in one-person households projected for the next two decades, suggesting that this will greatly dampen down the need for new housebuilding over this period. Yet it is worth noting that, according to the evidence brought together in Table 7.3, this pattern has been in place for at least two decades already. The projections to 2011, therefore, merely represent a continuation of a process which does not seem to have reduced the demand for owner-occupied housing significantly so far.

Related to this, there must be doubts about the wisdom of assuming that the type of demand for new housing construction, particularly in terms of dwelling size, can be read off directly from the type of information on projected net changes in household composition shown in Table 7.3. Matching 'small' households with 'small' dwellings is a process with which public-sector housing providers are familiar. A history of relatively generous standards for families under Parker Morris and the various bedroom standards has been mixed with very rigorous, even mean-minded, standards for small households. Table 7.4 compares the consumption of different sizes of dwelling by different types of household in 1988 for the South East of England. It demonstrates that only a small minority of one-person households occupy one-bedroomed properties in the owner-occupied sector but that over half do in social housing. It also shows that the occupation of small dwellings in the owner-occupied sector is heavily concentrated among young, particularly female, one-person households, for whom affordability and getting on the first rung of the housing ladder play an important part. There is little evidence from this type of analysis that future owner-occupier provision should be concentrated in small units, even though, given unchanging matching procedures, social-housing providers may continue to need to focus on smaller units.

These examples serve to illustrate the complexity of the relationship between demographic change and the housing market. In fact, the list could be extended. For instance, consideration of the matching process raises questions about the type, location and quality of existing housing being released by

Table 7.3 Household composition changes in South East England, 1961–2011 (thousands)

Household type	1961–71	1971–83	1983–91	1991–2001	2001–2011
Married couple households	+342	−219	−94	−161	−109
Lone parent households	+46	+177	+159	+181	+93
One person households	+420	+441	+345	+441	+421
Other households	−39	+145	+196	+133	+74
All households	+770	+542	+607	+594	+478

Sources: calculated from Department of the Environment data: 1961 and 1971 from headship dataset derived from Census of Population; 1983 from 1983–based household projections; 1991 onwards from 1989–based projections.

Table 7.4 Number of bedrooms by selected household types in South East England, 1988 (per cent)

Household type		1 bed	2 bed	3+ bed
One person households				
Single	Owner-occupier under 30 (male)	33	40	26
	Owner-occupier under 30 (female)	48	41	11
	Owner-occupier 30–44 (male)	14	48	38
	Owner-occupier 30–44 (female)	25	53	22
	All owner-occupiers	23	45	32
	Tenants in social housing	74	22	4
Widowed or divorced				
	All owner-occupiers	12	37	51
	Tenants in social housing	57	29	14
Married couple households				
	All owner-occupier	2	17	81
	All tenants	13	34	54

Note: Rows may not sum to exactly 100 because of rounding.
Source: Labour Force Survey 1988 (unpublished tables).

household dissolution and migration. Another major issue concerns the extent of housing in poor condition and the need for its replacement, which is likely to be of growing significance, possibly outstripping demographically induced demand in the longer term.

Enough has been said, however, to provide a basis for assessing in broad terms the likely scale of future housebuilding. In terms of overall building rates, there may well need to be a return to the situation of the 1960s and 1970s when new construction substantially exceeded the net increase in household numbers; for it is questionable both how long the reverse pattern of the 1980s (Figure 7.3) can be sustained and how long it is before more investment must be put into the replacement of poorer stock. As regards the progress of the 1960s baby boom cohorts, the evidence suggests that the impact on private housebuilding comes later and is spread over a longer period than household-formation figures alone suggest, extending well into the current decade.

Both these conclusions naturally have spatial implications over and above those allowed for in the county household projections presented in Figure 7.2. Moreover, they tend to aggravate the problems of the most pressurised parts of the country noted in the previous section. The 25–34 year olds who dominate the demand for owner-occupied housing are, during the 1990s, particularly well represented in the arcs of shire counties around London and the West Midlands conurbation. Meanwhile, the need for replacement housing can be expected to arise mainly in the large cities, but how far can it be assumed that this process will take place *in situ*? Will 'difficult to let' estates always be replaced with the same volume of housing on the same sites? It seems impossible in the long run. The location of new housing will need to reflect the new geography of employment and people's 'quality of life' aspirations. Ultimately, however, the outcome does not rest on market-place factors alone; it will also depend on political attitudes and the nature and effectiveness of planning policies.

Intervention in the future distribution of housing growth

It would be wrong to suppose that past development patterns are merely a reflection of demographic trends, even more so to assume that the current development process is merely expecting to respond passively to such trends in the future. The volume and impact of urbanisation pressures ensures that some form of land use policy intervention is often on the agenda. The scale of land consumption − around 8,400 hectares per annum for the national building rate of 200,000 in the mid 1980s at the prevailing density of 24 dwellings to the hectare − is such that, whether on 'greenfield' sites or in existing urban areas (and about 45 per cent of all housebuilding during that period was taking place on urban sites), the town and country planning system has to cope with the substantial conflicts of interest which are generated by development pressures. Few issues have triggered more interest in the environmental policy debate over the past decade. As yet, though, it is not at all clear what form the government response to these pressures will take over the next few years − a situation which arises not just from a reluctance to jettison the 'demographic bulldozer' viewpoint but also from weaknesses in the planning system and shifts in the direction of land use planning.

Planning mechanisms and policy shifts

The town and country planning system provides the vehicle for such land use policies and is one which involves both central and local government. There is, however, no effective national framework within which the future distribution of housing growth can be considered. Such centrally produced guidance as does exist has the function of guiding local planning authorities in the preparation of their plans and in the making of *ad hoc* decisions. A tentative framework of regional guidance, issued by the Secretary of State for the Environment, offers an outline policy framework for each region in the medium term (10−15 years), while Planning Policy Guidelines provide a sketchy policy framework and have a material, but generally partial, role to play in the formulation of local policies (see, for instance, DoE, 1992a, 1992b).

For most practical purposes it is the Structure and Local Plan system (jointly known as the Development Plan system) which provides the active framework for guiding the quantity and location of development in the medium term − usually ten years or less, by the time that plans become statutorily approved. Even now, however, some twenty years after the introduction of this version of the Development Plan System, coverage remains incomplete at the site-specific Local Authority District level because of a patchy commitment to the Local Plan system. The Development Plan system is in the process of being revised, requiring each Local Authority District to provide District-wide site-specific coverage (DoE, 1992c). This system is further supported by a requirement on each Local Planning Authority to demonstrate that a 5-year land supply exists for housebuilding at any given time, based on a rate which will ensure the fulfilment of the existing statutory Development Plan (DoE, 1992b).

Moreover, the political and ideological context within which this system is operated has shifted significantly since the beginning of the 1980s. In the early Thatcher years much of regional economic policy was dismantled, along with the new town policy. Key changes in government advice were also made which were seen as pro-development. Circulars in the early 1980s emphasised that in the planning system there should be a general presumption in favour of development (e.g. Circular 22/80, DoE, 1980a). This meant that, irrespective of the amount and disposition of planned development, other sites could be presented to the planning system by developers for active consideration if they could be seen to be reasonable in planning terms. At the same time, Circulars required Local Planning Authorities to ensure that there was always a 5-year supply of housebuilding land available (e.g. Circular 9/80, DoE, 1980b). These two factors were significant in encouraging a phase in the history of residential planning in this country which has become known as 'planning by appeal', since developers were not shy in coming forward with sites which met the above criteria in their judgement, but did not fit in with the plans (where they existed) of the Local Planning Authorities. Consequently the early 1980s was seen as a period in which the government was re-orientating the planning system along its 'ideological' lines (Goodchild, 1992; Rydin, 1986).

Following this early 'ideological' encouragement of development interests, the 'pragmatic' electoral/NIMBY pressures triggered a degree of retrenchment in the late 1980s which has carried on into the 1990s. Early radical proposals to restructure the Development Plan system have been watered down. An initial 'flexible' approach to development in Green Belts saw one of the earliest U-turns in the Conservative government's policy, which in the 1990s now wishes to be seen as the Green Belt's staunchest supporter, even at the expense of any pro-development wishes of the Local Planning Authorities. A similar U-turn has been experienced in terms of the early encouragement of new settlements, since when proposal after proposal has been rejected across the country.

Along with these U-turns, there has been a growing emphasis on development being channelled towards 'brownfield' rather than 'greenfield' sites. This will be reinforced by the relatively new concept of 'sustainable' development and sustainable cities, following the European Community lead (EC, 1990). While few, if any, practitioners argue against the merit of the beneficial re-use of derelict, unused or under-utilised urban (brown) land, there is a general concern that over-development of existing urban areas may result at the expense of the quality of life of those areas. As a consequence, this 'brownfield' site approach is increasingly referred to scathingly as 'town-cramming' (Fyson, 1992). All of these more recent shifts are now encapsulated in Planning Policy Guidance (DoE, 1992a and 1992b).

Current plans and proposals

These, then, are the planning mechanisms and the recent influences to which they have been exposed. The mechanisms themselves do not explain the levels of housing-land release incorporated in the plans. Virtually without exception, Local Planning Authorities are seeking to release less housing land over a given period than in the past. In many instances this is the result of the projected reduction in the level of net household increase outlined in Figure 7.1. However, in a growing number of localities, particularly in the south of England, the levels are based on assumed or implied reductions of net in-migration or even accelerating net out-migration. Linked to the NIMBYism referred to previously, many counties are claiming that they have reached (or are about to reach) their 'environmental capacity'. During the 1970s and early 1980s there were only two counties — Hertfordshire and Surrey, both heavily blanketed by the Metropolitan Green Belt — which appeared to be able to make such a claim legitimately. Because of the implementation of such a strategy, house completion rates in these two counties during the 1980s were relatively low, given their proximity to London, their environmental qualities, and their high post-war growth rates. These two counties have shown that plans based on a restraint policy can and do directly influence housebuilding rates. Demographically, the consequence is that Surrey experiences net out-migration and Hertfordshire meets indigenous growth only. Indeed, net out-migration is a normal consequence in those Districts which are embedded in the Metropolitan Green Belt, including, ironically, the former new towns of Harlow and Basildon in Essex, for instance.

The restraint-based planning philosophy of the Metropolitan Green Belt counties and districts of the 1970s has been increasingly extended to non-Green Belt counties in the south of England since the mid-1980s. Even those locations which have seen themselves as outside of the orbit of the London labour market and the general pattern of South East growth have found themselves under pressure from economic and demographic ripple effects. Together with locations on the receiving end of retirement pressures, they are beginning to question their hitherto relatively welcoming policy towards such growth. Few such locations have adopted policies recently which are as accepting of growth as they have been in the past. The majority of shire Planning Authorities are planning their housing land releases at levels which imply that net migration levels into their locality will have to reduce in the future.

For this strategy to be successful, much depends on the extent to which mobile population groups can be contained and catered for in the metropolitan cores and the economically disadvantaged regions. The housebuilding industry points to the impracticalities of such a strategy in the context of the very substantial long-term supply of land and existing vacant floorspace for industry, commercial and retail premises in the southern shires and London. Since there is no *planned* brake on job growth in the south generally, it is argued, the application of housing restraint policies across most of the south of the country makes little sense; if the large numbers of planned-for jobs were to materialise, then labour must inevitably be drawn towards them (HBF, 1988).

The early 1990s recession has inevitably complicated such an assessment, since jobs have been lost rather than gained and since labour mobility has been reduced to relatively low levels. Nevertheless, the pre-conditions for a paradoxical strategic planning situation still exist, waiting only for the next economic upturn to manifest themselves. The re-establishment of economic growth seems likely to continue to favour job growth in the South of England and to follow the general counter-urbanisation trends of the 1980s, reinforcing housing pressures in the southern shire counties.

Conclusions

The initial investigation of the 'demographic bulldozer' effect shows that a modest reduction in the rate of growth in the number of households can be expected nationally over the next two or three decades, but that the effect is neither as substantial nor as certain as many commentators have suggested in recent years.

Moreover, the locational consequences of this effect are not as simple as the national statistics might portray. The reduction in young adults and growth in the over 30s is unlikely to be spread evenly around the country. Some locations, such as the counties west of London, have a substantial propensity for future growth based on 'second generation' effects, i.e. growth within the local (indigenous) population. Others have, because of environmental, house price and economic growth reasons, a considerable propensity to attract in-migrants, particularly the coastal counties in the south of England. These propensities appear likely to remain strong in the future, subject to the vagaries of national economic performance. As a consequence, other locations may experience a decline in household numbers as households are 'pulled' towards those attractions and 'pushed' away from the poor performance of local economies and the unsatisfactory nature of housing and environment. Older urban areas, those with a legacy of traditional industry and pre-first world war terraced housing, and areas of system-built and high-rise public sector housing may be particularly vulnerable to this effect, not least because they have only a modest counterbalancing propensity for locally generated (indigenous) growth.

In terms of new housebuilding, changes in household numbers represent only part of the picture. Adjustments also need to be made within the system for mismatches in demand and supply relating to 'housing needs' and changing patterns of consumption of housing, for example, in terms of size, tenure and environmental context. Most of the adjustment patterns in recent decades point to a continuation of counter-urbanisation trends through the 1990s and beyond. In the longer term, the location of the ageing and deteriorating pre-first world war housing stock may have as significant a bearing on the volume and location of housebuilding as any demographic effects, much as it did in the 1950s and 1960s during the period of large-scale urban redevelopment. Whether the geography of poor condition housing corresponds to the geography of where people want to live in the future is a more problematic question.

The town and country planning system continues to see the containment of residential growth as a major, if not the major, strategic land use objective. The reasoning is two-fold. Firstly, the countryside is preserved and, secondly, infrastructure and service provision can be optimised and sustained, it is argued. However, three broad problems stemming from such a strategy have emerged during the 1980s. Firstly, the constraints which are imposed on those actively wishing, indeed needing, to live in rural environments are significant. Secondly, the adverse impact of the consequences of urban containment for the quality of life of those 'contained' residents — the 'town-cramming' effect — is seen as an increasingly important issue. Thirdly, the adaptability of existing urban housing provision and urban environment generally to changing aspirations and expectations of society is questionable. The popular belief is that the 'contained' housing and urban environment is deteriorating, counter to the steadily rising aspirations of society. The resolution of these social, political and planning conflicts will play a large part in determining the location of new housebuilding in the future.

Acknowledgement

Figure 7.2 is copyright of Anglia College Enterprises Ltd. All remaining figures and tables in this chapter use data which is Crown Copyright and is reproduced by permission of the Controller of HMSO.

8. GOING INTO A HOME:

Where can an elderly person choose?

Anne Corden and Ken Wright[1]

The growth of the elderly population is one of the most important developments of the second half of the twentieth century, as Chapter 1 has pointed out. Particularly important over the last two decades in Britain, as in most other European countries, has been the rapid increase in the numbers of people aged 75 years and over, and especially of the frail elderly. The proportion of the population that is aged 75 and over in the UK has grown steadily in the last three decades, up from 4.2 to 6.9 per cent in 1990. Over this period the absolute number of this age has grown by 1.75 million, or 78 per cent. Over half a million of this increase is accounted for by the 85+ group, with 866,000 people in 1990, over 150 per cent higher than 30 years earlier.

This phenomenon raises a number of major policy challenges. These include the problems of financing the mounting need for long-stay provision, of dealing with the intense pressures of acute illness, of providing domiciliary care facilities, and of developing a range of continuing care facilities for elderly people (Bosanquet and Grey, 1989). Against this background, it is perhaps not surprising that in the last few years central government has reacted by trying to curb the rate of growth in its financial commitment to long-term care provision, most notably through the National Health Service and Community Care Act 1990. The aim of current policy is to help elderly people remain in their own homes for as long as possible, with support and services. Moves into long-term institutional care become necessary, it is hoped, only when people reach a level of dependency where they can no longer be maintained within the community.

When a move does become necessary, a key problem is the availability of suitable accommodation. The expectation might be that for elderly people at the end of their lives, the move into long-term institutional care will be the last relocation they will make. The location of this final home is important to the quality of life of the elderly residents. Maintaining links with their original community will be easier if they have not moved far – it may be possible still to get to known local shops, their church or club; to keep the same GP; to go on enjoying the local paper and local radio; to be within a short distance of friends and neighbours. Very dependent residents may no longer be able to go out, but the location of the home will still be important because it will affect any visitors who might come to see them. Relatives and friends may themselves be frail elderly people, for whom the length and expense of the journey will be real considerations.

Quality of life may be affected in other ways, less immediately apparent but nonetheless important. The location of the home could be expected to affect accessibility of services such as hairdressing, chiropody and physio-therapy as well as the frequency of visits from professionals such as social workers. The staffing of homes is dependent on a good, local supply of labour; in particular, homes which are in remote rural areas may find it difficult to recruit suitable staff. There are financial implications; for instance, levels of fees vary between areas, and the range of services provided free of charge by different health authorities varies from one area to another.

Elderly people will thus have different needs and expectations about the location of the home into which they move, and we might hope that they have opportunities for exercising choice in this matter. Research findings from studies conducted by the authors at the University of York suggested that choice in location was very constrained for people moving into long-term care. Subsequent discovery that, contrary to assumptions, a move into a long-stay home was by no means a once-for-all move, and that there seemed to be a number of relocations after entry by apparently very old and frail people, raised considerable interest. Little is known about the locational process of moving into a home, and local authorities have become increasingly aware of the gaps in knowledge, particularly as they make plans for the new arrangements for community care which are being phased in for full implementation in 1993. This chapter is a first attempt to bring together what we know about the location process involved in moving into long-term care, the extent of current opportunities and constraints, and how these might change under the new arrangements.

This chapter is planned in four sections. The first examines the number of elderly people in institutional accommodation and the range of institutions to which they move. Secondly, we review the research evidence about the location process associated with movements into homes. In the third part of the chapter we ask how far the pattern of movements into care represents choice of location. Here, movements between homes after entry add a further perspective. Finally, we consider how moves into care, and choice of location, might change under the new arrangements.

Elderly people in long-term care

How many elderly people move into an institution? In fact, only a small minority are living in institutions at any time. The 1981 Census showed that only 3.1 per cent of elderly people aged 65 and over in England and Wales were not living in a private household, and this percentage had actually decreased since 1971. As would be expected, rates of institutional residence increase with age. A recent estimate is that in 1989, while only 1.2 per cent of people aged 65–74 in Britain were in some form of institutional care, the figure for people aged over 85 was as high as 23 per cent (Bosanquet, Laing and Propper, 1990). However, compared with other developed countries, including many of our European neighbours, our own rate of institutionalisation of elderly people is fairly low.

There are some regional differences: the 1981 Census showed the South East and the South West regions to have slightly raised rates of institutionalisation. Rates are also higher among women than men (Harrop and Grundy, 1991) and marital status and socio-economic factors such as tenure and social class are also associated with differential rates of institutionalisation (Grundy, 1989). The detailed results from the 1991 Census will bring these figures up to date, but meanwhile it has been suggested that by 1989 there were 29 per cent more elderly people in long-term care than would have been the case if age-specific rates had remained unchanged from the base year of 1981 (Bosanquet, Laing and Propper, 1990).

What sort of homes do elderly people move into? According to statistical returns from health and local authorities of places available in institutions, in 1991 there were just over half a million places (546,400) available in long-term care for the elderly, chronically sick and physically disabled people in the UK (Laing and Buisson, 1991). As Figure 8.1 shows, well over half of these places were in the independent sector. This includes private residential care homes run for profit; voluntary residential care homes, run by charities or religious foundations; and both private and voluntary nursing homes for more dependent people. There is still a substantial local authority residential sector, sometimes called Part III homes (Part IV in Scotland). In addition, there were 76,000 beds in NHS hospitals in 1991 in long-stay wards and cottage hospitals, almost equally divided between so called 'geriatric facilities' and 'psycho-geriatric facilities' for elderly people who are severely mentally deteriorated. There were also three experimental NHS nursing homes, which participated in an evaluation of public sector provision (Bond et al., 1989).

The important features of the present stock are the large proportion of residential and nursing homes developed over the last ten years and the way in which this provision has been financed. Between 1981 and 1990 the number of private and voluntary residential homes for elderly people and people with a physical handicap increased from 3,770 to 10,237 in England and Wales and the number of private and voluntary nursing homes increased from 1,078 in 1982 to 3,801 in 1990. At the same time the number of local authority residential homes reduced from 2,862 in 1981 to 2,745 in 1990 (Darton, Sutcliffe and Wright, 1993). The number of beds in geriatric medicine departments declined from 61,000 to 55,000 (Department of Health, 1991; Welsh Office, 1991). This means that the increasing demand for long-term social and nursing care has been met by the independent sector. Within the independent sector itself, the major increase in the supply of places has occurred in private homes. The number of private residential places in homes for elderly people and people with a physical handicap virtually quadrupled, up from 32,941 in 1981 to 126,267 in 1990, while the number of voluntary home places increased modestly from 26,900 to 27,478 over the same period (Darton, Sutcliffe and Wright, 1993).

The growth in private care has helped to equalise the geographical pattern

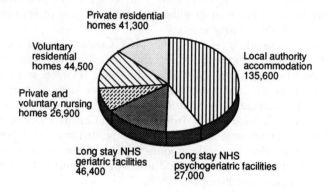

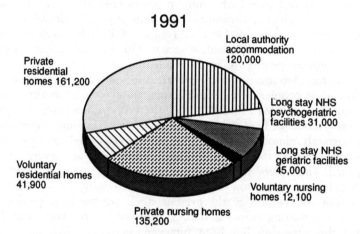

Fig. 8.1 Provision for places in long-term care for elderly, chronically ill and physically disabled people, UK, 1981 and 1991
(*Source*: based on data in Laing and Buisson, 1991).

of long-term care provision (Table 8.1). Traditionally, before 1980 there was a heavy concentration of private care in southern England, especially in the coastal retirement areas. Over the ten years 1978–1988, while the number of residents in independent residential homes in southern England increased by 136 per cent, the increase in northern England was 328 per cent, and in Yorkshire and Humberside it was 324 per cent. Although the gap in provision has narrowed over these years, the southern region of England still has the largest number of independent care home places (Hamnett and Mullings, 1992).

The spatial distribution of existing and new homes can be attributed to several demand factors. The general growth in the number of people over 75 years of age throughout the country has been an important factor (Larder,

123

Table 8.1 Number of people aged 65 and over in residential care, 1978−88, by region and provider

Department of Health region	% change in numbers 1978−88			% in private and voluntary homes	
	All homes	Local authority	Private and voluntary	1978	1988
Northern	+49.9	+4.1	+328.2	14.1	40.3
Yorkshire & Humberside	+51.9	−0.2	+324.1	16.1	44.9
North West	+65.9	−3.4	+295.6	23.2	55.3
Wales	+64.4	−0.7	+385.8	16.8	49.8
West Midlands	+58.8	−5.6	+333.7	19.0	51.8
East Midlands	+61.0	−1.1	+293.8	21.0	51.5
London North	+55.9	+1.1	+194.0	28.4	53.6
London	+2.8	−13.3	+34.9	33.5	43.9
Southern	+58.8	−10.1	+136.1	47.2	70.1
South West	+74.1	−7.2	+178.4	43.8	70.0
England and Wales	+51.5	−4.3	+188.0	28.9	55.1

Notes: Department of Health regions correspond to Standard Regions, except: 'South West' excludes Dorset and Wiltshire; 'London' is Greater London; 'London North' comprises East Anglia and all the South East as far south as Essex and Berkshire; 'Southern' is the remainder of the South East plus Dorset and Wiltshire.
Source: Hamnett and Mullings (1992), Tables 4, 5 and 6, and calculations derived therefrom.

Day and Klein, 1986) and the variation between local authorities in the percentage increase of elderly people partly explains the pattern of growth in private residential homes (Bochel, 1987). Economic factors such as low levels of unemployment and indicators of above average income and wealth in the community (e.g. proportion of households owning a car, owning a house, in social classes I and II) are associated with higher levels of private provision (Larder, Day and Klein, 1986). On the supply side, the availability of property suitable for conversion to care home use has been important to the growth of homes in some areas (Phillips and Vincent, 1988; Smith, 1986). In some cases local authority planning regulations have sought to restrict further expansion of the care home industry (Phillips and Vincent, 1988; Murray, 1985).

A considerable amount of the expansion in private care has been financed from the social security budget. In 1979 supplementary benefit expenditure on the support of residents in independent residential and nursing homes was only £10 million whereas for the year ending May 1991 annual expenditure on income support (previously supplementary benefit) reached £1,872 million (*Hansard*, 1991). This development has had significant effects on the choice of modes of care for elderly people. First, it has provided a perverse incentive for NHS and local authorities to opt for residential or nursing home care instead of pursuing the generally accepted objective of keeping elderly people in their own homes (Audit Commission, 1986a). Secondly it has encouraged the closure of freely provided or state subsidised places in hospitals or local authority homes and replaced them with facilities which are more expensive for a significant proportion of elderly people who are not eligible to receive income support to meet the charges made in independent homes. Against this it can be argued that elderly people now have a greater variety of choice in finding homes near to relatives or friends, and in finding a place which suits their own requirements. Both NHS and local authority facilities tend to have a relatively large capacity (40–80 places) and serve formally defined retainment areas within their own administrative boundaries. Independent homes on the other hand are smaller on the average (15–30 places), usually do not have formal rules about the geographical origins of residents and offer a variety of philosophies of care.

Geographical moves into institutions

Given the distribution of long-term care places described in the previous section, which elderly people go where, and why? This is a most complex issue that we are only just beginning to understand. It involves the various levels of dependency and different needs of elderly people (Wade, Sawyer and Bell, 1983); and socio-economic factors (Grundy, 1989); as well as the rationing, gatekeeping and advisory procedures that are operating (Hunter, McKeganey and Macpherson, 1988) and the knowledge base and exercise of choice by some elderly people themselves (Allen, Hogg and Peace, 1992). This section describes what we know about the geography of moves into homes. There are two main types of information available about geographical moves into institutions: large-scale analyses of migration patterns, using census data; and local cross-sectional surveys of residents in homes and of elderly people thinking about moving into a home. It is more difficult to find studies which monitor the process of moving into care for individual elderly people.

Migration patterns – large-scale analyses

Elderly people in general have relatively low migration rates. The 1981 Census showed only 5 per cent of people over 65 had lived at a different address a year earlier, compared with 10 per cent of the whole population. Among those recently retired and among very old people migration rates are higher than those of people in middle age. The former are associated with moves to traditional retirement areas, and the latter reflect the needs of very old frail people to move for services or care. Indeed, in 1980–81 the proportion of movers among elderly people aged 90 and over was higher than the proportion in any other 5-year age group in the whole population aged 55 and over (Grundy, 1987). Many of these moves of very old people are into institutions. Comparisons of retirement migration patterns and rates of institutionalisation help in discovering whether elderly people who move house on retirement are more likely eventually to go into an institution in the area to which they moved than the non-migrants. Service providers in areas of concentration of elderly people in the traditional retirement areas are of course very interested in answers to these questions.

Considerable work on the destinations and characteristics of elderly movers has been conducted by Grundy and colleagues, using the OPCS Longitudinal Study which enables identification of those elderly people who were living in private households in 1971 and had moved into institutions by 1981. Analysis showed that over one-third of those aged 75 or more in 1971 who had changed their address by 1981 were living in an institution at the later date (Harrop and Grundy, 1991). On the whole, people who had moved on retirement showed higher rates of institutionalisation than those who had not, but the rate declined with distance moved, and was lowest for those who moved furthest. Long-distance elderly migrants are healthier and socio-economically more advantaged than local or non-migrants (Grundy, 1987), so perhaps these advantages remain with them after retirement relocation, helping them to stay independent and avoid moving into an institution. However, rates of institutionalisation increased as the density of elderly people increased, in the traditional retirement areas. Grundy suggested that perhaps the high proportion and large numbers of elderly people in such areas had the effect of reducing the effectiveness of local community services, increasing the tendency to institutionalisation among elderly people already living in that area. During the decade 1971–81 some counties showed slight net gains of elderly people living in homes in 1981 who had been living in different regions of the county in 1971, particularly Somerset and Clywd. Other counties such as Devon and West Sussex showed net losses – more elderly people moved out of these counties into institutions than into the county and then into an institution.

The longitudinal analysis described relates to a period which precedes the massive expansion of the private long-term care sector of the 1980s. Similar analyses using 1991 Census data, once incorporated into the Longitudinal Study, will make it possible to look at the trends in migration and institutionalisation that accompanied the increases in care homes and the changes in the balance of provision.

Moves into institutions – local cross-sectional surveys

To complement the patterns and trends shown by national survey data, local surveys of residents and patients can be used to provide detailed information about geographical and locational aspects of moves into and between different types of institutions. In several recent studies questions have been asked about residents' previous addresses before admission to hospitals, local authority Part III homes, and the homes in the independent sector. We shall look now at each sector in turn.

Hospitals
Most hospitals serve local population areas and cross-sectional surveys of hospital patients in long-stay provision usually show few patients whose previous residence was far away. Somerset County Council's 1987 Census of elderly patients in long-stay hospital accommodation within the county found nearly all had previously been living within Somerset; a few people in hospitals situated close to the county boundary came from adjoining counties (Somerset County Council, 1989). An earlier survey of all patients in geriatric wards in Brighton Health District showed only 4 per cent had been admitted from addresses beyond the health district (Bennett, 1986).

Local authority residential care

In a similar way, local authority residential care accommodation is primarily provision for local people. The responsibility of local authorities, under Part III of the 1948 National Assistance Act, is to provide accommodation for 'persons who by reason of age, infirmity or any other circumstances are in need of care and attention which is not otherwise available to them' (Section 21). This responsibility extends to people 'normally resident within the area', although local authorities have powers to provide accommodation to people coming from further afield, and then have to negotiate with the local authority of origin in respect of financial arrangements. As well as assessment for need of entry to Part III homes, there is also a financial assessment of means to determine the resident's contribution to cost. In cross-sectional surveys of Part III residents, few are found who come from outside the local authority area. In the Brighton study described, only 3 per cent of residents in homes run by the local authority came from addresses beyond the Brighton Health District. The Somerset study showed 97 per cent of residents of Part III homes were living in Somerset prior to admission.

Independent sector homes

Private residential care and nursing homes are free to accept any residents for whom they are registered to provide care. Similarly, there are no official restrictions on the geographical areas from which voluntary homes may accept residents; although some voluntary associations, such as local charities or religious foundations, impose their own residence qualification for people seeking places. As a result, patterns of movement into the independent sector are not the same as into the public sector. More people move greater distances and across administrative boundaries.

Survey results demonstrate the extent of cross boundary moves into independent nursing homes. A census of nursing home patients in Harrogate Health District area in 1985 showed only just over half (57 per cent) had been resident in the Harrogate Health District prior to entry, and as many as 14 per cent came from addresses outside Yorkshire (Harrogate Health Authority, 1987). There were similar findings when the study was repeated in 1987 – 36 per cent of patients had addresses outside the Harrogate Health District when they previously lived independently. In the Somerset study described, 21 per cent of nursing home patients in 1987 were admitted from an address outside Somerset. However, examination of postal addresses showed that most of the immigrants had moved just across the county boundary from neighbouring counties, Avon, Devon, and Dorset.

Comparative data comes from a survey of a sample of nursing homes and their patients in eleven local authority areas in 1990 (Baldwin et al., 1991). Seventy-seven per cent of patients were over 70 years old. Proprietors were asked for each patient's previous home address, and their approximate length of residence there. Information was available for approximately 2,500 patients. While there are likely to have been some errors in reporting, the data provide a good general picture. Table 8.2 shows that overall, 14 per cent of patients had moved across a local authority boundary into their nursing home. However, some areas, for example Birmingham and Essex, had few direct immigrants.

Table 8.2 Patients moving into a nursing home from an address in a different local authority area

Local authority	No. of patients sampled	Percentage moving from different area
Birmingham	191	2
Cheshire	418	12
Dorset	231	13
Essex	105	9
Croydon	191	20
Nottinghamshire	270	17
Northumberland	179	23
Tyne and Wear	174	7
Shropshire	275	19
Clwyd	157	12
Gwent	306	11
Total	2,497	14

(*Source*: Baldwin *et al.*, 1991).

Of particular interest was the length of time those elderly people who had moved into a nursing home had been living in that area before entry. Table 8.3 presents data on 2,138 elderly patients who had moved into a nursing home in their own local authority area, and shows that three-quarters had lived at their previous address for more than 30 years! Some patterns of retirement migration are shown in Table 8.3. Dorset and Clywd, traditional retirement areas, have higher proportions of people who have moved into the area more recently. Overall, few elderly people had moved into a nursing home in their own local authority after a stay of less than one year in that area. The overall picture provided by these various studies is that, in the main, nursing homes serve fairly local populations, but do accommodate some long-distance movers.

Table 8.3 Patients moving into a nursing home in their own local authority area: length of residence at previous address (total number 2,138)

	Less than one year %	One to ten years %	11–30 years %	30+ years %
Birmingham	–	4	13	83
Cheshire	1	8	10	81
Dorset	6	22	20	52
Essex	–	2	2	96
Croydon	–	11	10	79
Nottinghamshire	1	13	22	64
Northumberland	–	7	–	93
Tyne and Wear	–	1	1	98
Shropshire	2	7	11	80
Clywd	–	25	19	56
Gwent	3	13	15	69
Total	1	10	14	75

(*Source*: Baldwin *et al.*, 1991).

The picture in independent residential care appears similar to that described for nursing homes. The Somerset study described showed 27 per cent of residents in independent residential care had moved into the county on admission to the care home. A study of 530 publicly funded elderly residents in private residential homes in four areas in 1986 suggested regional differences (Bradshaw and Gibbs, 1988). In that research the figure for Devon was similar to that described for Somerset, but in Sefton, an area of recent concentration of private homes, Bradshaw and Gibbs found a figure as high as 50 per cent, and in two other areas, Clywd and Lothian, lower levels of 13 per cent. This study found a similar scale of variation in the proportion of people admitted to homes who said they had moved to that area after retirement, ranging from 16 per cent in Lothian to 40 per cent in Devon. On average, people admitted from their own local area had lived there for 50 years (Bradshaw and Gibbs, 1988).

The studies mentioned so far have looked at individual local authority areas and health districts. Additional insights come from considering flows within a 'cluster' of adjoining administrative areas. Northumberland Social Services Department took this approach in a survey covering the five metro-politan districts in Tyne and Wear (Newcastle, North Tyneside, Gateshead, South Tyneside and Sunderland), County Durham and Northumberland – a coherent geographical unit centred on Tyne and Wear. Data were available on 3,100 publicly financed elderly people in 266 independent residential care homes, and 1,264 in 64 nursing homes, representing on average, 70 per cent of all residents in the homes surveyed (Corlett, Collinson and Pleace, 1991).

The Northumberland study showed that, of those residents receiving financial support from the Department of Social Security (DSS), around 21 per cent came from local authority areas other than those in which their care homes were located. However, a high proportion of movements between local authority areas was due to people moving quite short distances, into homes in another local authority area but still fairly near their previous address. There were no significant differences between the pattern of moves into residential care and moves into nursing home care. However, this study found a clear difference between the private and voluntary sectors. While 82 per cent of elderly residents in private homes came from an address in the same local authority area, this was the case for only 65 per cent in the smaller voluntary sector. The authors suggested that voluntary homes, which offered accommodation to people who shared religious, cultural or occupational backgrounds, attracted people from a wider area than the private homes. In one voluntary home in Northumberland 77 per cent of DSS supported resi-dents had come from outside that county (Corlett, Collinson and Pleace, 1991).

Taken together, the findings from local surveys suggest that the independent sector, both residential care and nursing homes, is serving local populations of elderly people who have been resident in the area for many years. Those localities in which homes contain higher proportions of newcomers tend to be areas of retirement destination or of recent rapid growth in private sector provision. In some parts of the voluntary sector there are distinctive patterns of location of origin among residents.

The hope would be that these patterns represent opportunities for choice for elderly people. The suggested model would be one in which the independent sector supplies alternative provision to that of the local authority, extending local facilities for elderly people, even those of limited financial means, who can choose a home in a known local area and maintain links with their community. At the same time, the independent sector would seem to offer the possibility of longer-distance moves in order to be closer to relatives or to return to pre-retirement areas of origin, or to choose a pleasant location at the seaside or in the countryside in which to spend the last years of life. This model would be a demonstration of some of the advantages of a mixed economy of provision, operating in the interests of elderly people and their families. The following section examines the extent to which these geographical moves do represent the exercise of choice by elderly people.

Choice of location

Moving into an independent home

The initial choice of whether to move into a home at all is constrained by the services available and their charges, the health and housing circumstances of the elderly people, and the availability of informal carers. Assessment studies on publicly financed entrants to independent residential care conducted in 1986 suggested that if services such as day care or sheltered housing had been available at the time of admission, 10 per cent of their sample could have remained in the community (Bradshaw and Gibbs, 1988). Interviews conducted in 1988 and 1989 with 103 elderly people who had moved into local authority and private residential homes during the previous 12 months also suggested that more intensive community services would have delayed or prevented entry to care (Allen, Hogg and Peace, 1992). More than a fifth of the residents felt they would have been able to carry on at home if more intensive home help or home care had been available. Around a quarter of the 74 people identified as the main carers of the sample of elderly people thought that the elderly person they cared for could have stayed at home with extra help.

Given that a decision has been made about entry to residential care, what evidence is there that elderly people consciously choose between the alternatives of local authority and private provision? A study of recently admitted residents of Part III and private homes in Suffolk (Phillips and Davies, 1990) suggested that, for the most part, it was not the residents who had made this choice. Generally it was a third party who influenced or determined the route into residential care – the route to the private sector was influenced by a relative; that to the public sector by a social worker. Allen, Hogg and Peace (1992) also considered that elderly people had little control over whether they went into a local authority or private home, and few had considered the alternative types of home.

In any case, the opportunities for choosing a location may be limited. The private sector is unevenly distributed (Corden, 1992). Within a county private homes tend to cluster in the main towns, and moreover, to concentrate in particular areas within towns (Smith, 1986). In order for there to be real choice of location for elderly people, easy access to information about homes would be a prerequisite. Lists of names and addresses of public and independent homes are available on request from registration authorities. Some independent homes advertise in local papers and Yellow Pages. Some local groups such as Age Concern and Community Health Councils try to keep up-to-date information about local homes but this is difficult since the independent sector is so volatile. The interviews conducted by Allen *et al.* with elderly residents and their carers confirmed that it was people's own local knowledge of the area that was important in deciding which home to move to. They remembered a home being built, or had watched it being converted or were familiar with its location through driving past it on local trips.

Further constraints on choice of location may be imposed by levels of fees, availability of vacancies and proprietors' own gatekeeping procedures. Fees vary widely between and within localities and sometimes people of limited means would have to scan a wide area to find an affordable price. More than 60 per cent of residents in independent homes claim income support towards fees (*Hansard*, 1990). Under the income support scheme in operation since 1985 there were standard national limits to the amount of benefit available. If an elderly person was unable to add further resources from personal savings, relatives or charities, their choice was limited to those homes with fees at or below the income support level. The Baldwin *et al.* (1991) survey of nursing homes in eleven local authority areas was conducted in 1990 when the current income support limit for elderly people in nursing homes was £200 and found that only 25 per cent of nursing homes surveyed quoted minimum weekly fees for new entrants of £200 or less.

If an elderly person is seeking a place to move into immediately (admissions to private homes are often responses to crisis situations), then choice of location will be limited to those homes with current vacancies. A survey of 352 independent homes in North Yorkshire in January 1988 (Corden, 1989) showed that nearly all vacancies in both residential care and nursing homes, at or below the basic benefit limits, were in the areas covered by the Harrogate and Scarborough offices – a long way from home for any elderly people living in the Dales, or the southern part of the county around Selby.

The final determinant in how far an elderly person can choose the location of a home is the proprietors' own gatekeeping policies. Proprietors of private homes make their own assessments about applicants' suitability. They consider whether the elderly person's needs are within the level of care which the home can provide and try to maintain a socially cohesive group (Todd, 1990). Incontinence, confusion or aggressive behaviour may restrict choice of location even further.

We have therefore identified two kinds of evidence about the locational process. The evidence from large-scale migration patterns, and from the home addresses of residents in cross-sectional surveys, suggests that most elderly people who go into long-term care move fairly locally. However, the geography of the market, and findings from detailed research on fees, vacancies, information available, proprietors' behaviour and elderly peoples' own accounts, suggest that eventual locations may reflect elderly peoples' real choice to a very limited extent. In order to find out how far the patterns observed represent actual preferences of elderly people or are the result of constraints, it is necessary to monitor elderly people as they approach the move into care and to follow the decision-making process and eventual location. Very little work of this kind has been attempted.

Moving between homes

Examination of moves between homes after entry casts further light on the degree of choice exercised by elderly people, as well as providing more information about the scale and dynamics of elderly population distribution. For a relatively high proportion of elderly people, the move into an institution from the community is not their last move.

We would expect to find many moves from hospital into residential care and nursing homes. Current emphasis on care in the community for elderly people, with reduction in NHS long-stay geriatric facilities, has led to hospital discharge policies that are highly dependent on availability of long-stay nursing home places in the independent sector. Thus most surveys of nursing home patients indicate a relatively high proportion have come to the nursing home from hospitals. These are often local transfers. Proprietors responding to the nursing home survey conducted by the authors in 1990 (Baldwin et al., 1991) reported that 50 per cent of their patients had been transferred from a hospital bed. However, there are difficulties in areas which do not have a well developed independent sector that can receive elderly people awaiting discharge from local hospitals. Concern has been expressed that lack of affordable vacancies in local areas has led to elderly people being discharged into nursing homes far away from their home areas and from their relatives and friends (HC, 1991a).

Some elderly people move in the other direction, of course, from nursing home or residential home to hospital. These are elderly people for whom independent proprietors are no longer able or willing to care. Bennett's survey of patients in the Brighton geriatric wards (Bennett, 1986) showed 7 per cent had been admitted from Part III, 3 per cent from independent residential homes and 1 per cent from nursing homes. These are mostly local moves, within health authority or local authority areas. We would also expect some transfers from residential care to nursing homes, again associated with increasing dependency and need for more intensive care. The authors' study of patients in nursing homes in 11 local authority areas (Baldwin et al., 1991) showed that, overall, as many as 9 per cent had been admitted from residential homes.

Moves also take place between homes registered to offer the same kind of care. Indeed, Allen, Hogg and Peace, (1992) noted how frequently elderly people had moved between residential care homes. Sixteen per cent of the private residents interviewed (but none of the local authority residents) had come to their home from another, in all cases except one from another private home. It was suggested that these moves were usually associated with choice and self-determination. According to Allen et al. (1992, p. 168) 'What is most interesting is the incidence with which elderly people moved from one home to another. This demonstrates the ease with which some private residents can exercise a choice and vote with their feet, an option that is more often than not, not open to local authority residents'.

Reasons for moving included wanting to be nearer relatives and dissatisfaction with the previous home. There were, however, some indicators that moves between homes do not always indicate personal choice. Some of the residents in the Allen, Hogg and Peace study had to leave previous homes when they closed down. There is also some evidence that elderly people have been asked to leave homes when they cannot meet fees (HC, 1991b).

Tracing moves between homes in a local area provides further information about the extent and direction of transfer. Using supplementary benefit records, all publicly financed residents in independent residential care and nursing homes in North Yorkshire and Somerset in 1985/86 were identified, and retraced after two years (Corden, 1988). There were 739 elderly people still alive and still in receipt of supplementary benefit in an independent home in early 1988. In this group as many as 14 per cent of the North Yorkshire residents and 10 per cent of those in Somerset had moved between homes within one local benefit office area. These were moves over short distances, for example from one side of Harrogate to another, so it is unlikely that they were moves to be nearer relatives. The directions of the moves between types of homes are shown in Table 8.4. In both counties, at least half of these moves were not in the direction that would be expected to be associated with increasing dependency – they were moves between homes of the same kind, or from nursing homes to residential care. Those who moved were very old people – in North Yorkshire more than one-quarter of the movers were over 90 years old!

Such findings raise more questions than they answer. Whose decision was it that such very old people who had already moved out of their own house into institutional care, possibly via a spell in hospital, then moved on to another, often very similar, home close by? Did this move represent choice, or necessity? What effect did such a move have on the elderly people's lives and those of their relatives? What were the financial implications for the public purse, since all these elderly people were receiving financial support? These questions are of prime importance in current policy development for

Table 8.4 Number of claimants aged over 60 in 1988 who had moved between homes in same local benefit office area, 1985–1988

	North Yorkshire	Somerset
residential home → nursing home	30	4
nursing home → residential home	6	2
residential home → residential home	25	6
nursing home → nursing home	9	–
other moves in independent sector, for example to or from dual registered homes, or unregistered homes	11	4
	81	16

(*Source*: Corden, 1988).

elderly people and underline the importance of understanding local population characteristics and dynamics; they need more detailed study.

Choice and future financial arrangements

Factors affecting choice

Before considering the implications for choice of location of the implementation of the National Health Service and Community Care Act 1990, it is useful to make some general points about the current situation.

Choice is affected by conditions of both supply and demand. On the supply side, choice of home depends on the availability of existing homes and of property which can be converted into long-stay accommodation. This has meant that in some localities homes are not always located near to the main concentrations of population (Corden, 1992). Furthermore, the public–private mix of provision is altering. Currently, the increased provision of private homes is being encouraged by financing arrangements which allow NHS and local authorities to use social security to finance the long-term care of elderly people. This has been welcomed by those who feel that competition in provision encourages efficiency and effectiveness provided that defined standards can be met. Others perceive it as a worry about caring for very frail, vulnerable elderly people away from the public eye (Wright, 1985). There is no evidence that standards of care in the private sector are lower than in the public sector. This debate is not the major concern here; the point is that shrinkage of the public sector is reducing choice.

The main demand side factors which affect choice are the prices that have to be paid for long-stay care and the income of the elderly people who enter it. The role of the public sector is crucial to the prices facing elderly people entering long-stay care. No charge is made for NHS care. Those people who have been transferred from, or are unable to obtain, an NHS bed face paying fees either to the local authority or to owners or managers of the independent homes. Some may be able to afford the fees from present income, others may have to use their accumulated life savings (including the capital in their house) to finance their care. Others are able to fall back on income support from social security to pay all or part of their fees. It is not surprising that there have been complaints from some people who have not been able to obtain or retain NHS care.

There is growing concern, too, that income support levels for people in the independent homes are not adequate to meet the costs of provision and that this is putting pressure on owners either to use cross-subsidisation of places within the homes or to seek 'topping-up' payments from relatives, charities or the statutory services (Association of Directors of Social Services, 1992). If these measures are not successful they may face closure. This in turn will affect choice. The higher rates of income support payable for nursing home care may encourage the use, and therefore the subsequent supply, of nursing homes compared with residential care homes (Westland, 1992). Choice, especially for the poorest elderly people, could become more restricted unless new financial arrangements can match care needs with the available supply.

The effect of the new arrangements

Following the National Health Service and Community Care Act 1990, changes take place in the arrangements for admission and financing of elderly people in the independent sector, with effect from April 1993. Under the new arrangements people of private means continue to pay their own fees and make private arrangements for admission. Elderly people of limited financial resources are to be assessed for care needs by the local authority, which also takes over the main financial responsibility for meeting that part of the fee that residents cannot meet. Income support on a similar basis to people in their own homes continues to be available to elderly people, plus a residential allowance, but the local authorities will now be looking for value for money in those independent homes with which they negotiate for places.

The needs of elderly people are to be carefully assessed. Since it is no longer possible to shunt the costs of residential or nursing home care on to the social security budget, it is intended that elderly people and their carers should have more support available to improve their chances of staying at home. The interesting question is how care managers can balance their need for value for money in the range of homes they use for placements with the variety of choices that can be offered to their clients.

The progress of elderly people is to be followed more closely and made visible for the first time. Care managers still have to consider continuity of care both in the original choice of home and in the transfer between homes as people's health state changes. Generally, all authorities may need help to cope with people who wish to move across administrative boundaries. An elderly person's home address and length of residence is likely to become a crucially important factor in the choice of possible location for their residential or nursing home place. The new financial arrangements both for the transfer of the relevant amount of the social security budget and for the local government general grant must be sensitive to this issue.

Conclusion

Despite the increased commitment to supporting elderly people at home, there is likely to be a continuing need for long-term residential and nursing provision for those who are very frail or ill. The aim must be to ensure that those who move into a home have as much choice and control in this process as possible. The location of the home is a key issue, affecting many aspects of residents' quality of life. The private market, which now dominates provision, has developed in a piecemeal way, with no overall planning to match provision to needs.

There is evidence that most elderly people move into a home situated in their local area, although some make long-distance moves in the independent sector. Transfer between homes is common, both between homes offering the same kind of care as well as between homes offering differing levels of care. We know surprisingly little about this locational process; how far moves represent choice or constraint. In theory, new arrangements should improve choice in location, but the practice may well be different. We need to know more about the expectations of elderly people at this stage in their lives, the effects of location on their quality of life, the extent of real choice of location available to them, the effects of the new legislation on the development of the market, and the effects of the geography of the market on wider issues such as patterns of retirement migration, local property and labour markets, and transport policies. There is scope for work at all levels – large-scale investigations by means of censuses and surveys, detailed monitoring in local areas, and qualitative approaches to help us understand the attitudes and expectations of elderly people and of those who provide their care.

Note

[1] This chapter draws on research commissioned by the Departments of Health and Social Security. The views expressed are those of the authors alone. This paper is Crown Copyright. Published by permission of the Controller of Her Majesty's Stationery Office. No part of this chapter may be reproduced in any form without the permission of the Department of Social Security.

9. HEALTHY INDICATIONS?

Applications of Census data in health care planning

John Mohan

It should be self-evident that health authorities would make use of Census information in the planning, delivery and evaluation of their services, yet arguably the potential of the Census has only recently begun to be explored. It is only in the last ten years that significant work has begun to be undertaken on questions such as the local uptake of services, localised variations in need for health care (Curtis and Woods, 1984), patterns of health service utilisation and their relationship to social need (DHSS, 1988) and so on. Not only do the characteristics of populations relevant to health care needs vary geographically at various scales, so too health service delivery is increasingly being organised on a local scale. Whereas some 30 years ago health care planning in Britain was dominated by the development of a national network of District General Hospitals (DGHs) to service populations of 100–150,000, today the focus is much more on identifying and acting upon inequalities within District Health Authorities (DHAs), with policy emphases being very much on locality planning and local delivery of services. This chapter reviews the use of Census data in the NHS in Britain, though referring largely to the position in England, and in particular the technical and political debates over the appropriate criteria for health service resource allocation. It then focuses on more local applications of Census data and argues that linkage to various other sources of information is essential for the Census to be used to its full potential. It concludes by summarising some of the potential pitfalls of the applications and technology discussed.

Geographical dimensions of factors affecting need for health care

It is not difficult to identify important spatial variations in factors which might affect either need for health care – for example, the distribution of the elderly, who consume substantially more health care resources than any other age group – or the ways in which services are planned – for example, social support networks without which effective community care may be impossible. More recently, there has been a growing interest in regional and local variations in the mortality experience of populations. A notable example would be the impact of ischaemic heart disease (IHD), the under-65 Standardised Mortality Ratios (SMRs) for which vary, in England, by a factor of 2 for men and a factor of 3 for women; for lung cancer, for women, the variability is by a factor of 6 (although this is partly a low numbers problem) (Chambers *et al.*, 1990).

The authoritative review of spatial variations in mortality recently published by OPCS (Britton, 1990) comprehensively demonstrates the variability in mortality between geographical areas in the UK. It confirms, for instance, the 'familiar regional gradient in mortality from high in the North and West to low in the South and East for both males and females' (Britton, 1990, p. 25). Using local authority areas, the report noted the frequency with which particular local authorities appeared in the top 20 SMRs for 25 selected causes of death; it found that there were 18 local authorities, predominantly around Liverpool and Manchester, but including seven London boroughs, for which the SMR for the selected causes of death appeared in this category. This report also examined a range of socio-economic and environmental influences on mortality, demonstrating links between the social class and tenure composition of authorities and their mortality experience. For example, those authorities with the lowest proportions of heads of household in social class I or II also experienced, in general, the highest SMRs; those local authorities where the proportion of households renting from the local authority was highest were also those where SMRs were greatest. As the report itself acknowledges, however, its use of local authority areas was 'outdated' and technical developments in the 1991 Census would permit more detailed analysis and also analysis for *ad hoc* spatial units.

At a more local scale, it has been shown that, even when areas of apparently similar social and economic conditions are compared, mortality experience varies substantially at the level of electoral wards. Here a technical difficulty has been that the OPCS's computerised death records were not postcoded until 1981 so longer-term studies of local mortality experience require the addition of postcodes and, even then, linking such data to wards in England is subject to a margin of error because postcode and ward boundaries do not match exactly. This kind of research demands the very careful linkage of mortality statistics to population data, because of the ways in which small numbers of deaths can have large effects on rates and ratios calculated for small populations, as shown by Townsend, Phillimore and Beattie (1988).

Some of the first work of this type was done in Scotland because of the ease of linkage of postcoded data to the Census, allowing interesting analyses of the relationship between mortality and deprivation (see Carstairs and Morris, 1991). A good recent example from England is the work of Phillimore (1990). He notes that a problem bedevilling small-area studies of mortality variations has been 'the absence of an obviously suitable population denominator at small-area level in years well away from the Census' (p. 373). His study of localised variations in mortality in Middlesbrough and Sunderland had therefore to rely on the careful use of ward-based population estimates for the calculation of SMRs for small areas, based on electoral wards, of apparently similar socio-economic status. He was able to show that the mortality experience of Middlesbrough was around 60 per cent worse than the national average, while that of Sunderland was 30 per cent above the national average; what was interesting about this was the unusual *similarity* in social conditions between the two localities.

These observations, and the kind of inequalities summarised here, raise important questions: even if links can be established between social conditions and variations in health status or mortality, just how are they to be interpreted? And if no explanations can be found through aggregate social indicator, what other sorts of information may be required to make sense of the kind of variability described in this case? Furthermore, how can health authorities best use the data available in the Census for decision-making? In this latter context, there are probably four obvious applications of the Census (Knight, 1992):

- helping authorities know their markets, by providing information on the distribution of relevant population groups. Variations in these will have considerable financial implications
- stimulating debate on how effective health service policies have been: what changes have taken place in health status and how far did/could the health service influence them?
- informing resource allocation decisions, particularly by identifying local areas with specific problems, possibly by using data on the distribution of social deprivation
- assessing the location of points of service delivery *vis à vis* the distribution of the population.

The value of the Census for these purposes will depend not just on the information it contains but the use made of it, so the rest of this chapter focuses on the potential and some of the pitfalls of using Census data in the NHS.

Social factors and Census data in NHS resource allocation

Census data and national resource allocation policies

Prior to the 1970s there was little evidence of any extensive or sophisticated use of Census information in the NHS in England, beyond the kind of norms-based planning evident in the 1962 Hospital Plan, which envisaged distributing hospital capacity in accordance with national standardised beds/population ratios. Incremental adjustments to the existing pattern of services were therefore the order of the day. This had at least as much to do with technical limitations (availability of computing facilities and machine-readable data) as with wider philosophical issues (the absence of an agreed theory of need). The search for a more robust way of allocating funds began in the late 1960s during Richard Crossman's tenure at the DHSS, and was continued in the 1970s, producing the Resource Allocation Working Party (RAWP) formula in 1976. Designed, in principle, to offer a way of allocating funds to Regional Health Authorities (RHAs) in accordance with needs, the formula was subject to considerable criticism, and it is worth spelling out some of the points made.

The original RAWP formula was population-based, modified to take account of age and sex composition and also of SMRs as a proxy for morbidity. There was no explicit allowance for social factors in the formula. By comparing existing levels of service provision with 'target' levels of provision, RHAs (and their constituent health authorities) were classified as being either above or below target in terms of funding. The way RAWP was to be implemented was by a steady relative redistribution of funds from what were regarded as 'overfunded' RHAs (principally, but not always exclusively, the Thames regions) and, within them, inner city health authorities where population had declined leaving behind hospital provision which was regarded as excessive in relation to the size of the population; the locational inertia of the teaching hospitals was not unconnected with this. Perhaps it was not surprising that it was from these locations that arguments were advanced in favour of including a deprivation factor in resource allocation formulae: prominent hospital consultants advocated this in the letter columns of *The Times* and the medical press, while local authorities and MPs from inner London were equally convinced that additional funds were required to meet London's special needs. Nevertheless the suspicion remained that the deprivation arguments concealed a defence of the status quo — the existing distribution of resources — by those with an interest in its maintenance, especially in view of the limited evidence presented about exactly how social deprivation was translated into need for health care.

The problem with including a deprivation weighting into resource allocation calculations was that deprivation indices were highly correlated with indicators of supply of services, so that simply demonstrating some association between needs indicators and levels of utilisation of services was not sufficient proof that deprivation translated into greater need for health care. Three positions were advanced in the debates on this point (which are reviewed by Mays and Bevan, 1987). Firstly, social factors were said to place greater demands on hospital services because deprivation produced greater morbidity; secondly, utilisation rates were said to be high because, in the absence of good primary care in urban areas, the population make use of hospitals as an alternative, a pattern of utilisation which may be judged inappropriate; and thirdly, utilisation rates were said to be high because supply of services creates its own demand (in the absence of any financial barriers to access to health care) and so, irrespective of the quality of primary care, residents of areas with high levels of hospital provision will use hospital services.

The DHSS's review of RAWP attempted to tackle this question by deriving measures of social deprivation and access to hospital services, and analysing the intercorrelations between them. The evidence received from the various RHAs, while perhaps not surprising, showed the range of factors the 14 regions considered relevant to the determination of need for health care, which generally reflected special pleading on behalf of the jurisdictions of the RHAs involved. Thus those with large inner city populations argued for deprivation weightings and indices of the presence of ethnic minorities while rural areas retaliated with arguments for sparsity weightings to reflect both the social costs related to the isolation of their residents and the greater costs of running dispersed patterns of services. The review of the formula advocated a deprivation weighting as well as various other modifications, although this conclusion was attacked (see below).

Only six months after this conclusion was reached (DHSS, 1988), however, this proposal was apparently swept aside in the review of the NHS. The revised system was essentially population-based, with adjustments for the demographic composition of the 14 RHAs; however, SMRs were eventually incorporated in the formula. The new allocations were used to allocate funds for the purchasing of health care by the 14 RHAs and so did not reflect historic patterns of use of services which had had such an effect on the regional and subregional allocations of funds under RAWP. Despite the weight of the case argued for deprivation factors by the RAWP review, the White Paper made no mention of it and argued instead for simplicity and transparency in resourcing the NHS. While consistent with the government's wider social philosophies — after all, if there was no such thing as society, but only individuals and their families, as Mrs Thatcher stated, then one would not expect a resource allocation procedure to include much other than population counts — technical sophistication was clearly jettisoned. If there was a technical argument against a deprivation factor, it was presumably that differences in deprivation were not an issue at the regional level; it was arguably at the sub-RHA level that such factors become more important (see also Mays and Bevan, 1987).

The White Paper laid great stress on the importance of the RHAs planning their services and allocating resources in accordance with local needs, this being indicative of wider decentralisation trends in public-sector planning. This became apparent in the different ways in which RHAs chose to incorporate socio-economic information into their subregional formulae. A briefing from the National Association of Health Authorities and Trusts (NAHAT, 1992) summarises the ways in which RHAs chose to do this. It is notable that, while most RHAs chose to use the national formula set by the DoH, there were differences in some criteria used. For instance, SE Thames did not — initially, at least — include a deprivation weighting in its formula, which meant that inner London DHAs with ostensibly similar levels of deprivation to those elsewhere in the capital were treated differently in the allocation of funds. Nor was there consistency in the indicators chosen. For instance, NW Thames used a composite of the Jarman score, the results of a regional morbidity survey, and the SMR for those aged under 75, whereas the SW Thames RHA excluded any SMR weighting and used a composite social deprivation weighting instead (they had previously used an ACORN (A Classification of Residential Neighbourhoods) — based system for subregional RAWP). While arguably this results from health authorities taking decisions which relate more closely to local circumstances, it does seem inconsistent to have a particular deprivation factor included one side of an RHA boundary and not on the other. At the RHA scale this may not matter too much (since there has been considerable convergence between RHAs on target levels of funding under RAWP) but more locally, it can be highly significant, as a discussion of the use of deprivation indices in primary care makes clear in the next section.

The interregional and intra-regional distribution of resources may well be affected by the proposal to incorporate a measure of long-standing illness or disability in the 1991 Census. As with other Census indicators this will have to be considered carefully and its implications for resource allocation worked out. In particular the relationship between the incidence of long-standing illness, need for hospital care, and utilisation will have to be carefully assessed before decisions are taken on whether and how to incorporate this measure in national formulae. In the short term it may well be that it will be used more for within-authority decisions on the disposition of resources, as is argued later in the chapter.

Indices of deprivation or workload? The Jarman index and
deprivation payments to GPs

There has been considerable controversy over the ways in which deprivation might be measured and used as a criterion for health service resource allocation. One of the best examples is provided by the debates over the use of the Jarman index of deprivation in allocating payments to General Practitioners (GPs). Jarman, an inner London GP and Professor of Primary Care, has been arguing for several years that social deprivation generates additional workload for GPs, on the basis of findings of a survey of the factors which GPs thought affected their workload to which 1,802 GPs responded. This produced ten Census indicators, whittled down to eight in the final version, weighted according to the average score assigned to that variable on a scale from 0 to 9 when asked to rank them 'according to the degree to which it increases workload or contributes to the pressure of work when present' (Jarman, 1983, p. 1,706); thus the variable 'elderly living alone' received a weighting of 6.62 in calculating the index, while that for ethnic minorities (people born in the New Commonwealth or Pakistan) received a weighting of 2.5.

The Jarman index has received criticism because of its method of construction and because its utility as an indicator of health need has been questioned. Among the criticisms summarised by Senior are the following: the subjectivity of the choice of variables and weightings used in the index (which applies also to the obvious competitors, such as the Department of the Environment's Index or the Townsend index); whether the index measures the risk or reality of deprivation (the Townsend index, by contrast, focuses squarely on measures of material deprivation such as overcrowding and unemployment, and some of those included in the Jarman score (e.g. the elderly living alone) may not be deprived; whether using indicators in this way really measures GP workload or the impact of social deprivation upon workload (for example, those who change their address frequently cause extra administrative work for GPs but this is not necessarily anything to do with deprivation); and, finally, the skewness of the values for components of the index may have the effect of giving undue weight to the most positively skewed variables – which just happen to be those given least weight by GPs – thus unintentionally treating some localities more favourably than others. Finally, virtually any composite index of deprivation has the problem that its components are probably intercorrelated in some way so, according to Senior, 'it is very difficult to explain to policymakers exactly what these indexes are measuring' (Senior, 1991, p. 84). These debates are also reviewed by Carstairs and Morris (1991), who stress the absence of indices of deprivation designed specifically for the identification of health problems and the various weaknesses of the numerous composite indices proposed, at various times, by researchers seeking to relate health inequalities to deprivation.

It is the adoption of the Jarman index as a means of allocating funds – deprivation payments – to GPs that has raised particular controversy. Since April 1990 GPs have been eligible to receive a deprivation payment for each patient resident in a 1981 Census ward identified as 'deprived' according to Jarman's UPA score. Payments rise steeply at thresholds in the UPA score of 30, 40 and 50. No payments are made for patients in a ward classed as not deprived (UPA under 30); the payments are then £5.05 for someone in a 'low deprivation ward' (UPA between 30 and 39.99), £6.65 in a 'medium deprivation ward' (UPA between 40 and 49.99) and £8.80 per patient in a high deprivation ward (UPA over 50).

Senior (1991) has calculated the consequences of moving to a deprivation payment system based on the Townsend index of deprivation. This is a simpler indicator, which includes only four variables (unemployment; people in overcrowded households; households with no car; and households which are not owner-occupiers). It is less subject to the problems of skewness which bedevil the Jarman index, and also focuses more sharply on material aspects of deprivation. Broadly speaking, the Townsend index will redistribute re- sources away from inner city FHSAs with high proportions of ethnic minorities in their populations, towards 'northern' FHSAs, especially those with high proportions of people living on overspill estates on urban fringes. Good examples of gainers would include Merseyside, Tyne and Wear, Humberside and Cleveland; losers under the Townsend index include several FHSAs in London, such as Lambeth, Haringey, Ealing and Waltham Forest. The sums are not trivial: GPs in Liverpool FHSA would gain a total of £1 million if the Townsend index were used (Senior, 1991, table 6).

Even when these problems are ignored, there are problems with the system used for calculating the payments to GPs, which demonstrate some of the limitations of Census data when used in this way. The Census is infrequently updated and so does not respond to changing conditions, although of course the Census covers the whole country. More seriously, Census data refer usually to standard units – wards, most commonly – and the distribution of wards categorised as 'deprived' may not correspond to the distribution of deprivation. Therefore it is important to 'treat Census measures of deprivation as neither precise nor accurate' (Senior, 1991, p. 85) but this is not the case with the present system of deprivation payments. The present system can lead to substantial variations; a case in point would be GPs in adjacent wards whose UPA scores were just above and just below 30, in which case the former GP, assuming an average list size of 2,000 patients, would be £10,000 per annum better off. An obvious point is that deprivation does not occur in discrete stages as the DoH's formula suggests; it is a continuous variable. Senior therefore recommended a tapered application of deprivation payments in which scores would start at a much lower level of the UPA score but would rise by 10 pence per unit increase in the score. This would reflect more realistically the continuous gradation of deprivation and remove artificial boundaries between wards categorised as being of low, medium and high deprivation, but it would also involve the politically unacceptable admission that larger areas of the country were regarded as 'deprived'.

A still more sophisticated development would be the use of EDs as the geographical basis for payments, but while they can permit finer geographical variations in 'deprivation' to be detected, this innovation would depend on whether or not patient postcodes can be accurately located by ED. There is a proposal to produce a directory of whole and part postcode units by 1991 ED, which may be acceptable depending on how serious problems of part-postcode units proves to be (Senior, 1992, p. 5). Ultimately these debates are not going to be resolved by more sophisticated refinements of the existing formulae, however desirable; the more relevant question is whether or not Census indicators, as bases for resource allocation, actually measure the things they are supposed to measure, such as workload. Carr-Hill and Sheldon (1991, p. 395) suggest that what is needed is a detailed evaluation of the deprivation payments system in terms of whether they affect issues like list size, access to GP services, and practice income and expenditure.

In summary, the basic problem shown both by the debates on hospital and community health service resource allocation and on the use of deprivation weightings in primary care is that of confusion over just what indicators are supposed to be measuring. Without replaying long-running debates over just what constitutes 'need' (Davies, 1968), composite social indicators are not necessarily accurate predictors of need, even though associations may well exist with areas of ill-health. While they are of some value as indications of relative 'need' and therefore of use to health authorities in their new purchasing role (in which the assessment of needs is central), they do have limitations in this regard; this is why some commentators have argued that 'needs assessment needs assessment'! (Gabbay and Stevens, 1991). This implies that such indicators need to be linked to local surveys of morbidity, or complemented with data on the proportion of people reporting long-standing illness or disability, as will be possible when the 1991 Census results are released.

Management applications: locality planning, locality profiles and geographical information systems

Locality planning and locality profiles

As well as resource allocation between authorities there are a number of potential applications of Census data at a local scale, in managing health services within authorities and in responding to local variations in need, which require much more detailed management information. Moves towards locality planning reflect the interaction of three trends. Firstly there has been a move away from a concern with regional inequalities in the distribution of resources to a focus on more local issues. This results from the progress made in the redistribution of resources regionally and the recognition of the persistence of significant localised inequalities, some of which appear anomalous, in health and mortality. Since the Black Report (Townsend and Davidson, 1982) there has been rapid expansion in the number of studies of local variations in mortality (summarised, for instance, by Whitehead, 1987). Secondly, policy with regard to the provision of health services is much less capital-led than formerly; it has swung away from concentration on acute surgical and medical treatment and towards decentralised, community-based services, following recommendations of key reports in the mid-1980s (e.g. Cumberlege, 1986) and reflecting wider dissatisfaction with institution-alised facilities and interventionist medicine and surgery. Thirdly, with the encouragement of organisations such as the WHO, and with a recognition that health inequalities are not soluble solely by health services, there has emerged an intersectoral approach to health problems in which collaboration between all relevant agencies is the goal. For some (e.g. Davies, 1987) this denotes a move away from a role for health authorities simply as providers of health care towards a more coordinative role: their function is now to pull together all the diverse strands of resources available in a locality, with a view to a coordinated attack on health problems and inequalities.

This requires the definition of localities for organisational and resource allocation purposes (see also Kivell, Turton and Dawson, 1990; Curtis and Taket, 1989). Most localities to date have been organised around operational boundaries in efforts to achieve greater coordination between local authority and DHA operational units. To be effective locality planning requires accurate knowledge of variations in social conditions within localities, and so one can expect a growth industry in the production of locality profiles. Census data will be especially widely used for targeting purposes and for identifying populations at risk. Just as marketing agencies focus on locations with a particular socio-demographic profile, so one might expect health authorities to wish to concentrate on particular locations in their resource allocation decisions. Jones and Moon (1987) report an illustration of the use of Census data for preventive health campaigning, showing how literature concerned with advising the elderly on the prevention of hypothermia was targeted on specific EDs within the authority; because of the geographical concentration of people aged over 75 living in poor quality housing, it was possible to reach all of this group by concentrating on only 36 per cent of the EDs in the health authority.

An extension of such techniques is the use of composite social indicators derived from the Census for classifying the types of neighbourhoods within health authorities, and there is some evidence of enthusiasm for this within the NHS (e.g. Knight, 1992). However, some of the techniques for doing so were developed primarily for market analysis purposes, such as ACORN. Their theoretical basis is limited (usually some form of cluster analysis), and the range of variables included is not necessarily always relevant to health care planning, so they may be of doubtful value although they have been used in subregional resource allocation formulae (Hale, 1991) and for the identifi-cation of linkages between the incidence of specific conditions (e.g. food poisoning) and particular socio-economic profiles (Brown, Hirschfield and Batey, 1991). What is really required is a purpose-built classification of Census data relevant to a health service context.

GIS and health service planning: Technology in search of datasets?

Arguably health care planning cannot rely on Census data alone: there are no unambiguous indicators of need for health care and it is the relationships between Census information and other sources of data − principally on provision of, accessibility to and use of services − that is crucial. Consequently it is essential that Census data be linked, at an appropriate scale, to information about utilisation of services and about patterns of provision. Here some examples are given of the applications of Geographical Information Systems (GIS) technology and some cautionary warnings are pointed out. Note that there is confusion over what is meant by GIS: Maguire (1991) defines GIS in terms of map processing, database management, and spatial analysis; GIS is more commonly confused with computer-assisted cartography which, despite emphasis on quality cartographic display, lacks data retrieval and analysis capabilities.

The potential of this technology was recognised some time ago (e.g. Mohan and Maguire, 1985) but it is only recently, driven by both technical and political developments, that GIS technology has begun to find applications in the health service. Many of the technical problems which impeded development of GIS in a health care context are beginning to be ironed out. These include the lack of spatial referencing for routinely collected data (resolved by the recommendations of the Korner Committee on Health Services Information (DHSS, 1985) that all health activity information should be postcoded), the absence of digital boundary information (for EDs, wards and health authority areas), and the inability of software packages to link together different data sets. In principle there is no reason why health activity information should not now be linked to Census data in a routine manner. The political reasons concern the general pressures on health authorities, as with other public agencies, to make more 'efficient' use of available resources, and information technology has been recognised as crucial in this regard. In addition, since the NHS White Paper, changes in administrative arrangements in the NHS have emphasised the need for health authorities, in their new roles as purchasers of services, to take an active role in researching the needs of their jurisdictions for health services; this has implied a greater concern with analysing the variability of local needs and developing appropriate responses.

However, even when NHS data are postcoded, aggregating unit postcodes into postcode sectors produces areal units with boundaries which cut across those of zones used in collecting census data. A consequence is that point-in-polygon techniques have had to be used to associate postcodes with areal units such as census wards (Gatrell, Dunn and Boyle, 1991). This means using the Central Postcode Directory to associate each unit postcode with a 100-metre grid reference. Because of the ways in which this is done, there are certain inaccuracies inherent in it, notably the problem that the grid reference may fall outside the area covered by the unit postcode in question. In addition there have been efforts, led by marketing companies, to attach 1-metre grid references to each property in the UK, and while intended primarily for marketing purposes this clearly would be of use in a health care context. Averaging of the coordinates for the individual properties comprising a unit postcode would produce a centroid for that postcode which could then be linked to ward or ED-level population data. Gatrell, Dunn and Boyle, (1991) describe the extent of the inaccuracies which can arise in this context and discuss strategies for minimising the problem. This situation will improve to some degree with the 1991 Census since the digitising of ED boundaries will permit more accurate matching of postcoded data (and indeed any data to which grid references are attached) with Census information, admittedly at considerable computational cost. OPCS will also produce a directory of whole postcode units and part postcode units listed by ED, including the number of usually resident households in each postcode/part postcode unit, which will also help overcome that problem to some degree. According to OPCS, this directory will allow assignment of postcodes to EDs and to pseudo-EDs (whole unit postcodes which most nearly approximate EDs). Where unit postcodes are split between two or more EDs, they are assigned to the ED which has the highest number of households from that postcode. The inaccuracies arising, which will be more limited, could only be overcome by producing the entire Census for spatial units aggregated entirely from postcodes in some way.

Despite these problems GIS are spreading within the NHS, with Gould (1992) reporting that 72 per cent of DHAs in England and Wales currently make some use of some type of computer software (GIS or computer-assisted cartography (CAC)) for handling geographically based information. However he also noted some definitional confusion over just what a GIS was; many respondents confused GIS with CAC. Applications largely included health needs assessment, epidemiological research, health profiles, and analysis of variations in use of services. Gatrell and Naumann (1992) in fact question whether full-blown GIS are really required by health authorities, arguing that for most requirements (such as annual public health reports) computer mapping systems suffice. At the RHA level they contend that some capability above and beyond a computer mapping system may be needed in view of the RHA's strategic role, for instance in reviewing the location relative to the RHA's population of major health facilities. By linking together (using the ARC/INFO proprietary GIS) data on population distribution, hospital locations, road networks, and estimated journey speeds, they assessed what proportions of SE Thames RHA's population lived outwith a 30 minute drivetime (at peak times) of a hospital accident and emergency department. They point out the utility of this kind of analysis in simulations of the possible effects of major closures or changes of use of health service facilities, and argue that increased sophistication will be possible with the 1991 Census when ED boundaries are available. Despite the occasional example of this kind of work (and others reported by Brown, Hirschfield and Batey, 1991), relatively little spatial analysis and spatial modelling has been undertaken using GIS in the NHS, a point echoed by Gatrell, Dunn and Boyle, (1991); for most DHAs GIS is essentially equated with computer mapping.

Mortality, morbidity and need: linking the Census to local and national health surveys

In the absence of agreed measures of health care needs, mortality has often been used as a surrogate for data on morbidity, as has been the case, for example, in regional and subregional resource allocation formulae. In the absence of reliable and frequent data on health status, health authorities have perforce had to rely on local morbidity surveys in order to gather information, using such instruments as the Nottingham Health Profile (NHP) for self-reported morbidity, or local health and lifestyle surveys to gather data on the prevalence of factors thought likely to impact on health status (such as smoking). This has been expensive and has meant that the effort devoted to it has varied among health authorities, although a number of authorities have conducted their own NHP-based studies (Curtis and Woods, 1984) or health and lifestyle surveys (e.g. NW Thames RHA). Moon and Twigg (1988) report a survey of attitudes to nutrition issues in Portsmouth, in which the approximately 1,260 respondents' addresses were attached, via a postcode-ED file, to their appropriate EDs. The social characteristics of the EDs had previously been defined using a cluster-analysis based typology, and statistical associations were analysed between the type of area in which individuals lived and their attitudes to nutritional questions, using the chi-square statistic. Few of the resulting statistics proved significant, which meant that the area classification could not be used as a surrogate for the survey data. However inspection of contingency tables showed interesting variations between areas of different social status in terms of willingness (for instance) to cut down on particular components of diet, thus highlighting what sort of areas should be prioritised in health education campaigning. Linking of survey and Census data is a task fraught with difficulties; there are important methodological and technical problems associated with linking individual and aggregate data (e.g. Curtis, 1990). Inferences about population characteristics drawn from aggregate data are subject to the ecological fallacy, so prediction of any aspect of health status or health behaviour from Census data, even where extensive survey work has been undertaken, is fraught with problems. Nevertheless, Curtis (1990) claimed that indicators such as the UPA score may be a good predictor of variations in the proportion of a population reporting illness and service use, once demographic variations are taken into account. However she observed that differences in reported illness and service use are only *statistically* significant for extreme variations in the UPA index score, and recommended caution in using it as a basis for resource allocation decisions except where there were major variations in the index. (Note that multi-level modelling may represent one way out of this impasse; see Jones and Moon, 1990).

The most useful single development in the 1991 Census, from the point of view of health care managers, is the inclusion of a question on the extent of self-reported illness and disability. Previous Censuses have not included such information, though they did include information on the proportions of the population who were permanently sick, and the national General Household Survey (GHS) contains too few households or individuals for disaggregation to the district health authority level. There are, of course, several ways of asking about limiting long-standing illness and disability. Research about the views of academics on the proposed 1991 Census suggested that of four possibilities canvassed (chronic illness affecting activities; disability hindering employment; ability to care for self/mobility; and acute illness in the previous two weeks) all had different uses in different contexts, though the first would offer comparability with the GHS. The 1991 Census question concerns whether a person 'has any long-term illness, health problem or handicap which limits his/her daily activities or the work he/she can do' and should, in principle, provide comprehensive information on at least the extent of such problems in the community. In addition, information to be made available on dependant populations and carers will be of value in community care planning. However, there remains no question on indicators of income and receipt of benefits, despite this being the single most important topic identified by a pre-Census survey of academic social scientists (Marsh *et al.*, 1988); this might have helped in the identification of links between indicators of ill-health and poverty. Other sources of data on incomes, such as local authority housing benefit registers, may not be easily accessible or available to health service users without collaboration and may be subject to data protection restrictions.

Concluding comments

To reiterate a point made at the outset, the competitive conditions under which health authorities now operate, plus political pressures such as demands from the 'Citizen's Charter' for agencies to demonstrate the effectiveness of what they are doing, necessitate the close investigation of health needs. This will demand the integration of a range of data sources, including those on social and economic conditions, mortality, health service utilisation, local health surveys, and so on, but the range of potential sources which could be interrogated should not allow agencies to be blind to some rather basic questions.

Perhaps a fundamental question to be answered by users of the Census in a health planning context is just what is being measured when Census indicators are used and therefore just what implications this has for planning. Do variations in socio-economic conditions imply variations in health service resource allocation procedures (for instance in the amounts allocated for treatment or preventive care) or are they variations which cannot be dealt with by the health services (for instance, the effects on the health service of conditions caused or exacerbated by living in poor housing)? Even the limited long-standing illness question, though helpful, is not unproblematic – its relationships to other indicators of ill-health and mortality need to be established. Composite indicators of social conditions, such as those originally developed in commercial geo-demographic agencies, also need to be used with caution. Their theoretical basis is weak and their links to health need questionable.

There will never be some absolute index which provides all the answers; at best, Census data, even when linked with other sources of information, will only indicate relative levels of need (shares of the cake) but not what the size of the cake should be. They will nevertheless raise questions about the distribution of the cake; if localities of apparently similar socio-economic status appear to have substantially different levels of resource utilisation; this should raise questions in the minds of managers. However, Carr-Hill (1990) argues that the complexity of the post-White Paper NHS (with funds being allocated to fundholding GPs as well as to purchasing agencies, possibly on the basis of quite different criteria), plus the difficulties of using Census data to determine 'need' for health care, will be such as to force the 'DoH to give up and simply hand out resources on a capitation basis – if one can any longer believe Census data after the introduction of the Poll Tax' (p. 201).

In similar vein, questions may be raised about the use of GIS in health care planning. While technologically sophisticated in some cases (and note that some 'GIS' have been little more than statistical packages linked to a graphics package), these arguably have yet to deliver the results of which they may be capable (with exceptions which are beyond the scope of this chapter such as Openshaw's Geographical Analysis Machine (Openshaw et al., 1988) for analysing the pattern of point datasets such as individual cases of childhood leukaemias). 'Technology in search of datasets' is one comment which springs readily to mind. It may be, however, that GIS applications will in practice be relatively routine, linking together sources of information for management purposes rather than providing path-breaking insights, although some are beginning to argue for an incorporation of greater analytical capability into existing GIS (e.g. Gatrell, Dunn and Boyle, 1991).

It is interesting to note, finally, that the provenance of many of the technical developments discussed here lay in the desire of the Conservative government of the 1980s to introduce its particular version of a managerial agenda for the service, in line with broader economic goals of restraining state expenditures. It would be ironic if one consequence of these techniques was that they revealed considerable evidence of unmet need and of under-utilisation of services, so producing demands for increased expenditure. A more plausible conclusion, however, would be that the greater amounts of information that are becoming available will simply allow an avowedly limited cake to be shared out with a greater degree of apparent technocratic rationality.

Acknowledgements

I am very grateful to Tony Gatrell, Myles Gould, Philip Kivell, Graham Moon, and Martyn Senior, who kindly sent copies of recent papers and/or work in press; none of them are in any way responsible for what I have made of their work, however. Graham Moon also kindly commented on a previous draft as did the editor of this volume.

10. MAKING WAVES?

The contribution of ethnic minorities to local demography

Vaughan Robinson

As other authors in this volume have made clear, the last 30 years have seen remarkable changes in the characteristics and distribution of the British people. None of these changes, however, has been more significant than the growth of the black and Asian minorities, with their distinctive spatial distributions and internal demographics. The settlement of these groups in numbers has had a profound effect on certain English cities whilst passing by other parts of England, Wales and Scotland. Whilst some streets in northern mill towns were already devoid of white residents by the end of the 1970s and one ward of Ealing (the Northcote ward) already had a population which was 85 per cent black or Asian by 1981, the whole of Wales was still estimated only to have 14,800 black or Asian residents in 1988 (Haskey, 1991).

The aim of this chapter is to examine the scale and significance of the contribution of ethnic minorities to local demography in Britain. After outlining briefly the context for black settlement, the chapter comprises four sections. The first discusses the growth of the ethnic population from its pre-war roots, through the period of mass migration and into the phase of family reunion or creation in the late 1960s, 1970s or early 1980s. Central to this discussion is the concept of waves of immigration since different groups arrived at different times to face different circumstances, so the next section looks at the influences which helped to steer the various groups to particular destinations. The third section then describes, in detail, the geographical distribution of the main ethnic groups and the extent to which the patterns of the early 1970s have persisted subsequently. This reveals how certain regions, counties, cities, towns and parts of towns have experienced ethnic minority settlement whilst others did not, and therefore demonstrates how vital it is to study specific localities since these are so different. In the final section, two particular issues are used to illustrate the potential significance of uneven ethnic distribution.

Setting the context

Britain has until very recently been a country receptive to international immigration, with a reputation for accepting both the persecuted (see Panayi, 1992) and those seeking their fortunes. The Italians, the Jews and the Irish, to name but three groups, all came to this country in sizeable numbers from the late 1800s onwards. Once here they often retained their distinctiveness both in behaviour and cultural mores, and this mapped itself onto the British cities in which they settled. Their distribution was highly spatially selective with only certain cities and certain neighbourhoods proving attractive (Pooley, 1977; Waterman and Kosmin, 1986). Thus, ethnic groups have been consistently important elements in the local demography of some cities since Victorian days.

Whilst some might argue that what separates the period since the 1950s from this earlier era is the number of immigrants involved, this is not really true. What has been far more significant has been the transition from white immigrants to those of colour, since this has invoked unparalleled xenophobia, prejudice, stereotyping and discrimination. Fear, ignorance and colonial stereotypes (Rex, 1970) have combined into a complex of racial exclusionism. Despite legislation to outlaw some forms of racial discrimination – passed in the 1965 and 1976 Race Relations Acts – there is still a huge weight of evidence that points to covert and overt discrimination and the attitudes which underly it (see e.g. Robinson, 1987; Brown, 1984; Phillips, 1986; Brennan and McGeevor, 1987; Commission for Racial Equality, 1990). One cannot ignore the conclusion that blacks and Asians are denied access to scarce resources and equal opportunities.

The growth of the non-white population

One of the difficulties facing those researching ethnic minorities in Britain is the inadequacy of official data. The national censuses have until recently relied upon birthplace as a surrogate for ethnicity, but in Britain's case there are two problems with such a substitution. First, because of Britain's colonial past, over 100,000 children were born to white government employees in the Indian subcontinent. In the Census these children are categorised as Indian even though they are British in ethnicity. Secondly, as some of the earliest groups to arrive now reach demographic maturity an increasing proportion of their number have been born in the UK. Over half those of Afro-Caribbean descent in the UK were born here: in official statistics based on birthplace, their ethnic origins are lost and they are grouped with white Britons. More recently, other national data sources have changed over to definitions based on self-ascribed ethnicity and from 1991 onwards so has the national Census, radically improving the prospects of accurate research.

Most of the literature on Britain's visible minorities focuses on the period from the 1950s onwards, when mass migration began. However this undervalues the importance of the period before the Second World War, for it was often then that the main ports of entry were established through which later arrivals flowed. Subsequent changes in the mode of transportation and the numbers involved certainly changed the pattern of settlement but in many cases communities remain in these early ports of entry as relic features in the ethnic geography of the country. Fryer (1984) and Visram (1986) both provide excellent accounts of the early presence of blacks and Asians respectively and they describe the significance of seafarers in the earliest phase of settlement. Seaports were their main destinations, a factor which helps account for the early and, in some cases, continuing importance of Cardiff, Bristol, South Shields, Glasgow, London and Liverpool in patterns of settlement. The same is also true of the Chinese who began to arrive in Britain from the mid-nineteenth century onwards but who started to develop settled communities around the turn of the century in ports such as Liverpool, Cardiff, and the East End of London (Robinson, 1992a). Thus it is important to remember that whilst some cities in the UK will have experienced settlement by visible minorities only in the last twenty years, others will have histories of settlement which go back almost a century.

Table 10.1 focuses on the post-war period and demonstrates how the ethnic minority population (including those of mixed parentage, Africans and Arabs) has grown considerably since 1951. Moreover it is likely that the 1951 and 1961 figures derived from birthplace statistics considerably overstate the size of the real non-white population, which was probably below 100,000 in 1951 and nearer 350,000 in 1961. If this is true, the ethnic minority population of the UK has grown by 25 times in 35 years.

The source of this growth has also changed over time. Increasingly growth has come from natural increase not from immigration. Whereas net ethnic minority immigration totalled 115,000 in 1961, this had fallen to 24,000 by 1986. In parallel, the contribution made by natural increase has grown from 55 per cent in 1971 to 71 per cent in 1986.

Disaggregating this total reveals that there have been waves of migration at different times. West Indians were the first to arrive, in the mid 1950s, followed by Indians (late 1950s through to the 1970s), Pakistanis (late 1960s and 1970s) and latterly Bangladeshis (1970s). Table 10.2 demonstrates how the different groups have matured demographically in this same order. Of course, these temporal trends have local spatial implications. Areas of early West Indian settlement will now be experiencing reduced fertility and population ageing, whilst areas of Bangladeshi settlement will still be recording rising fertility and very youthful populations.

Table 10.1   Size of ethnic minority population in UK, 1951−90

	000s	% of population	increase p.a. 000s
1951	200	0.4	
1961	500	1.0	30
1971	1,200	2.3	70
1976	1,612	2.9	82
1981	2,100	3.9	98
1983−5	2,350	4.3	83
1988−90	2,624	4.8	55

Sources: All data Shaw 1988 except 1976 (Immigrant Statistics Unit, 1986) and 1988−90 (*Population Trends*, Vol. 67, 1992).

Table 10.2 The demographic imprint of successive waves of ethnic minority immigration to the UK

	Total Period Fertility rate		No. of births (000s)			% aged <16 yrs	% change in population
	1981	1987	1971	1976	1987/90	1985-7	1981-87/9
West Indian	2.0	1.9	12.5	7.1	3.8	25	-9
Indian	3.1	2.7	13.3	12.0	8.6	31	7
Pakistani	6.5	5.2	8.2	8.1	12.9	43	52
Bangladeshi				1.4	4.8	50	116

Source: % change figures from 'In brief', *Population Trends* 63 (1991). Births: Pakistani and Bangladeshi figures are for 1987, others 1990.

The process of immigration and its spatial implications

The phasing of immigration is closely related to its spatial impact, because the destinations of the newcomers were strongly influenced by the distribution of available jobs at the time of entry. There is evidence that specific employers or industrial sectors sought to recruit ethnic labour in the late 1950s and early 1960s: these arrangements had clear spatial implications since they led to marked concentrations of blacks or Asians in particular plants or in regions where particular industries were concentrated. We know that the London-based British Hotels and Restaurants Association recruited West Indian male and female labour, as did London Transport and the large teaching hospitals in London. We also know that the London Brick Company in Bedford attracted West Indian men to replace their predominantly Italian labour force and that Wolf Rubber in Southall recruited Asian labour. Each of these arrangements produced nuclei of spatial and sectoral concentration which can still be traced through to the present day.

Those migrants who came without pre-arranged employment were faced with a different pattern of economic opportunities according to the wave of migration of which they were a part. For these people, the location of pre-war ports of entry would have been very important since they could provide a stock of cheap accommodation and access to information networks. Social and kin networks would then operate to locate and acquire housing and suitable employment. As a result, early cores such as Brixton became central to West Indian settlement and numbers there grew from a handful in 1948 to 5,000 in 1955 (Patterson, 1963): today Brixton is still one of the main cores of Afro-Caribbean settlement. On a different scale the Nevisian community of Leeds developed in response to the actions of three individuals who provided shelter for fellow islanders on their arrival in the UK (Byron, 1991).

More broadly, though, the pattern of West Indian settlement was formed by the availability of employment at the time: in the foundries and engineering shops of the West Midlands, in the car plants of London, Oxford, Coventry and Birmingham, in the paper industry of High Wycombe, and in the transport undertakings of Greater London and its surrounding satellites. The beginning of the Indian wave of immigration was only slightly behind the

West Indian wave so the earliest Indians were able to take advantage of some of the same employment niches in the same parts of the country. The subsequent vintages of Indian migration and more particularly the later Pakistani wave had to find alternative occupational or spatial opportunities, often located in the more distant regions of the country. Municipal public transport undertakings in places such as Manchester, Oldham, Preston and Blackburn were enthusiastic employers as were British Rail. Northern engineering and metal companies in cities such as Sheffield also took on Asian labour.

However perhaps the most significant difference to earlier waves was the growth of opportunities within the textile and woollen industries of Lancashire and Yorkshire and the clothing and footwear industries of the East Midlands. In both regions employers were driving down labour costs through capital investment, a degradation of work practices and stagnant wage levels. Asian migrants were quick to occupy the vacant niche created by the withdrawal of white male labour and by the legal restrictions imposed on the employment of white female labour, and in many cases they made particular textile mills effectively their own. Again, social and kin networks were vital in the acquisition of information about cheap housing and new vacancies (see Anwar's 1979 work on Rochdale for example), and ethnic work gangs with unofficial middlemen were equally important in ensuring that fellow Pakistanis and Indians were recruited when work did become available. Again this combination of economic and temporal factors had a spatial outcome, with major nuclei of Pakistani and Indian populations developing in towns and cities such as Leeds, Bradford, Huddersfield, Manchester, Blackburn, Preston, Oldham, Bolton, Leicester, Nottingham and Dundee.

Finally, one can contrast the concentrated patterns of black and Asian spatial distribution with the very dispersed nature of Chinese settlement. The latter opted for a different occupational niche, namely the restaurant and take-away trade, which positively required dispersal in order to prevent saturation of local markets (see Robinson, 1922a).

Geographical distribution of the ethnic minority population

The spatial outcome of these processes can be examined at a number of geographical scales ranging from the broad regional picture down to the level of individual neighbourhoods and streets. This section describes these patterns and assesses the extent to which they reflect the distribution of original settlement by the newcomers.

Table 10.3 shows the distribution of West Indians, Indians, Pakistanis and Bangladeshis between the Standard Regions of England and Wales and how it has changed over the past two decades. The data reveal how, as the different waves of migrants arrived, they had to adopt progressively more diversified patterns of distribution to find vacant employment and housing opportunities and how, in general, they had to move further and further from the South East of the country. As the first wave of arrivals, the West Indians were the least evenly distributed in 1971, in particular being more heavily concentrated in the South East of the country than other groups. The South East was also important for Indians, but less so: they looked further to the

Table 10.3 Regional distribution of ethnic minority groups in England and Wales, 1971–88 (%)

Region	West Indian			Indian		
	1971	1981	1986/8	1971	1981	1986/8
North	0.3	0.2	0.4	0.8	1.3	0.6
Yorkshire & Humberside	5.0	4.9	4.9	5.9	6.5	5.7
North West	4.6	4.8	5.0	6.6	8.3	7.3
East Midlands	4.9	5.2	3.3	7.4	9.4	11.3
West Midlands	15.2	15.6	17.0	20.9	20.8	19.0
East Anglia	1.1	1.0	0.7	1.0	1.2	0.8
South East	65.3	64.9	65.5	53.8	48.4	53.8
South West	3.0	2.7	2.7	3.3	3.0	1.4
Wales	0.6	0.6	0.6	0.4	1.0	0.6

Region	Pakistani			Bangladeshi		
	1971	1981	1986/8	1971	1981	1986/8
North	2.0	2.2	3.4	–	0.0	6.1
Yorkshire & Humberside	19.7	21.1	21.1	–	6.6	5.3
North West	14.6	15.7	18.3	–	13.5	4.6
East Midlands	3.5	4.2	3.7	–	11.9	3.3
West Midlands	20.9	21.7	20.1	–	9.1	16.6
East Anglia	1.2	1.4	2.6	–	0.0	0.3
South East	35.6	31.1	29.1	–	58.0	62.0
South West	1.5	1.6	0.2	–	0.0	1.3
Wales	1.1	1.1	1.3	–	0.8	1.1

Note: columns may not sum to exactly 100.0 because of rounding.
Source: 1971 and 1981 figures from national censuses; 1986/8 from Haskey (1991) *Population Trends* 63; 1981 Bangladeshi figures are unpublished OPCS estimates (see text). No Bangladeshi figures available for 1971.

north for opportunities and found these in the West Midlands, East Midlands and North West. By the time the Pakistani wave arrived, they were forced to look even more to the north, with the effect that their centre of gravity moved further north still, such that the South East was only about half as important for them as it was for West Indians. In contrast both the North West and Yorkshire became major foci of Pakistani settlement. This leaves the paradox of the Bangladeshis who might also have been expected to locate themselves in the northern industrial regions of the UK or even in the more peripheral industrial areas such as Wales where competition would be even less. One can only speculate as to why such a pattern of settlement did not evolve, but it seems likely that the Bangladeshis suffered from the heavy deindustrialisation of the UK in the period after the 1973 oil crisis and in the Thatcher years: the vacant niches which had been available in the 1950s and 1960s simply did not exist in the mid and late 1970s as the regional economies crumbled and thousands of manufacturing jobs disappeared. For the Bangladeshis, unemployment in the South East was perhaps preferable to unemployment in South Wales.

Table 10.3 also allows the examination of the extent to which these patterns of settlement have persisted since the early 1970s. It shows that West Indian regional concentration has been remarkably constant, with a strengthening of the West Midlands' position being the only notable change. The Indian distribution also appears relatively static, with the main trend being towards a greater concentration in the East Midlands. The Pakistani pattern, however, looks to have been changing, with a substantial fall in the South East's share and consistent growth in the North, Yorkshire and the North West: the group's centre of gravity has thus been moving further north over time. The Bangladeshi figures confirm the growing importance of the South East and West Midlands and suggest also that the North West and East Midlands have experienced considerable declines in their relative significance. The 1981 Bangladeshi figures are unpublished internal OPCS estimates and are subject to a large sampling error, which means that they should be treated with caution.

Significant concentration also occurs within the Standard Regions. Around six out of every ten Afro-Caribbeans live in Greater London and eight out of ten live in the main conurbations (see Table 10.4). London is less important for Indians but is attracting more of them as time passes, as indeed is the West Midlands conurbation. In contrast, both of these centres are losing their relative share of the Pakistani population, which appears to be increasingly attracted to Greater Manchester. The Bangladeshi minority in 1986/8 had a distribution more akin to that of the Afro-Caribbeans, albeit with a third

Table 10.4   The conurban distribution of ethnic minority groups in England and Wales, 1961–1988 (percentage of totals for England and Wales)

Ethnic group	Greater London	West Midlands	Greater Manchester	Merseyside	Tyne & Wear	West Yorkshire	Total
West Indians							
1961	57.3	14.4	3.2	0.6	0.2	3.0	78.7
1971	55.2	13.0	3.4	0.5	0.1	3.4	75.6
1981	57.0	12.8	3.4	0.4	0.1	3.0	76.7
1986/8	58.4	15.6	4.0	0.1	0.1	3.5	81.7
Indians							
1961	33.5	11.0	3.2	1.7	1.4	3.7	54.5
1971	33.5	14.0	3.9	0.7	0.6	5.0	57.7
1981	36.3	16.5	3.9	0.6	0.6	4.6	62.5
1986/8	42.6	16.7	4.6	0.4	0.3	4.9	69.5
Pakistanis							
1961	21.7	24.1	4.0	0.8	1.6	19.3	71.5
1971	22.0	16.9	9.0	0.3	0.6	15.9	64.7
1981	19.6	18.6	9.3	0.2	0.7	17.3	65.7
1986/8	17.5	17.1	11.4	0.1	1.4	19.4	66.9
Bangladeshis							
1981	45.2	12.9	8.5	0.7	1.0	5.0	73.3
1986/8	47.8	15.8	3.3	0.7	5.2	4.8	77.6

Sources: 1961 data from Census Birthplace and Nationality Tables; 1971 data from Lomas (1973) using birthplace and head of household's birthplace; 1981 data based on CRE estimates using birthplace and head of household's birthplace; 1986/8 data from Haskey (1991) using self-ascribed ethnicity.

concentration in the Tyneside conurbation where there are few other ethnic minority group members. Again, the 1981 estimates have a high sampling error and should perhaps only be used to indicate the growing importance of the Greater London and the West Midlands.

Figures 10.1a to 1d take the description down to a finer spatial scale by employing counties, London boroughs and metropolitan districts for 1986/8. Data are mapped as Location Quotients where values of 1 indicate a similar relative concentration to that of the white population, values of more than 1 indicate ethnic minority overrepresentation, and values of less than 1 denote ethnic underrepresentation relative to whites. Figure 10.1a confirms the importance of the main conurbations for West Indians. Figure 10.1b adds Leicestershire to the Indian conurban distribution. Figure 10.1c illustrates the importance of non-conurban areas to the Pakistanis, with significant numbers being found in the towns of Lancashire, Berkshire and Staffordshire. Figure 10.1d confirms the importance of a broad swathe of counties around Greater London to Bangladeshis, as well as confirming the importance of that city itself. The maps also highlight the fact that several areas are important settlement foci for a number of different ethnic minority groups. Greater London, West Midlands and West Yorkshire are important to all four groups. Greater Manchester is important for three groups, whilst Nottinghamshire, Bedfordshire, Berkshire, Hampshire, Leicestershire and Kent all represent significant areas of settlement for two groups.

Figure 10.2 provides a more detailed consideration of the spatial overlap of the three largest groups through the use of a ternary diagram and data for those individual London boroughs, metropolitan counties, and counties which had more than 5 per cent of their 1986/8 populations of ethnic minority stock. It reveals a surprisingly clear picture. The inner London boroughs which have relatively high ethnic minority proportions of their populations tend to be dominated by West Indian settlement with this group forming between 60 and 80 per cent of the ethnic population: Indians usually form the majority of the remainder. In outer London, the pattern is reversed with Indians typically forming between 60 and 75 per cent of the ethnic population with the balance made up of West Indians (20−30 per cent) and Pakistanis (10 per cent). In the other major conurbation, the West Midlands, over-lap between groups is much more in evidence with towns often being 25−35 per cent West Indian, 45−60 per cent Indian and 15−25 per cent Pakistani. The urban centres of West Yorkshire stand out as a third group, dominated by their Pakistani populations (50−70 per cent) and with a minority of Indians (20−30 per cent) − a pattern repeated by the Lancashire textile towns, which is not surprising given that their employment structure and housing stock closely mirrors that of their Yorkshire counterparts. Finally, there are the outliers from the major clusters. Leicestershire stands out as a major centre of Indian settlement, Bedfordshire is a county with a very mixed ethnic minority population and Bolton and Tameside represent towns with a mix of the two Asian groups.

Figure 10.3 adds a temporal dimension to the analysis by looking at the growth in ethnic minority populations within the same centres between 1981 and 1986/8. If Figures 10.1 and 10.3 are considered together they demonstrate

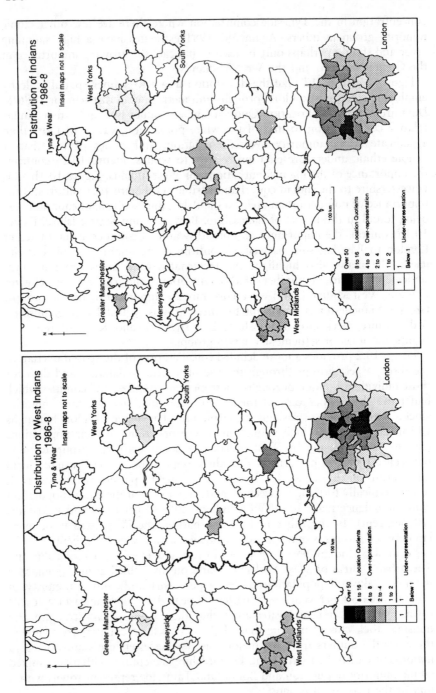

Fig. 10.1 The spatial distribution of Britain's ethnic minority populations, 1986/8 (See text for definition of Location Quotient).

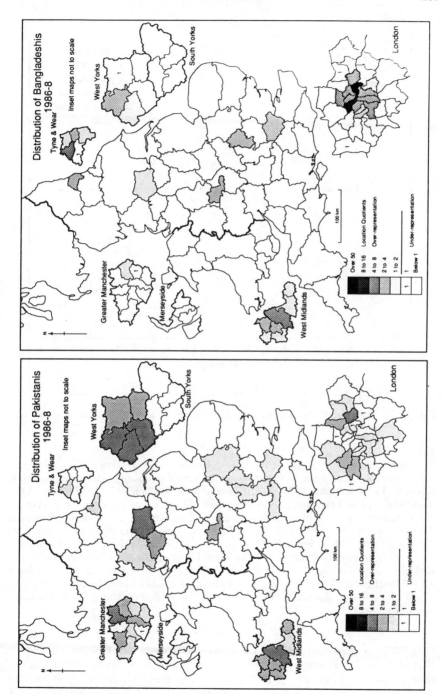

Fig. 10.1 (continued)

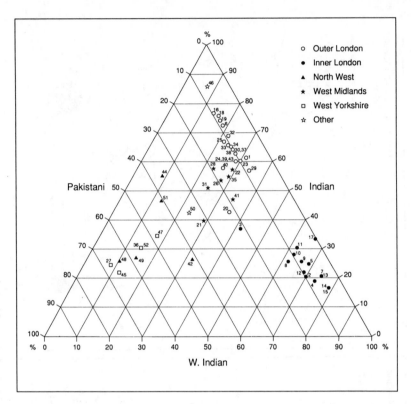

Fig. 10.2 The spatial overlap in the distribution of Indians, Pakistanis and West Indians, 1986/8.

Key
1 = Brent, 2 = Tower Hamlets, 3 = Newham, 4 = Hackney, 5 = Haringey, 6 = Ealing, 7 = Lambeth, 8 = Wandsworth, 9 = Westminster, 10 = Kensington & Chelsea, 11 = Camden, 12 = Hammersmith & Fulham, 13 = Islington, 14 = Southwark, 15 = Lewisham, 16 = Hounslow, 17 = City of London, 18 = Harrow, 19 = Barnet, 20 = Waltham Forest, 21 = Birmingham, 22 = Wolverhampton, 23 = Croydon, 24 = Merton, 25 = Redbridge, 26 = Sandwell, 27 = Bradford, 28 = Coventry, 29 = Enfield, 30 = Greenwich, 31 = Walsall, 32 = Hillingdon, 33 = Kingston-on-Thames, 34 = Richmond-upon-Thames, 35 = Dudley, 36 = Kirklees, 37 = Sutton, 38 = Bexley, 39 = Bromley, 40 = Barking, 41 = Solihull, 42 = Manchester, 43 = Havering, 44 = Bolton, 45 = Calderdale, 46 = Leicestershire, 47 = Leeds, 48 = Rochdale, 49 = Oldham, 50 = Bedfordshire, 51 = Tameside, 52 = Wakefield

that areas of predominantly West Indian settlement are relatively stagnant, that areas with a strong Indian presence are more likely to be expanding and that centres of Pakistani population are the most likely to be expanding rapidly: this is entirely in keeping with the timing of their arrival in the UK and also continues trends apparent in the 1970s (see Robinson, 1989).

Perhaps of greater importance than the mere fact of spatial concentration itself is the nature of the places involved. Are they the new sunrise settlements associated with services and information technology or do they comprise the rust-belt cities of declining opportunity? This has been a relatively neglected aspect of ethnic minority settlement in Britain, only two authors having considered the twin issues of the black and Asian urban hierarchy and urban typology in any depth (Jones, 1978; Robinson, 1986; Robinson, 1989). Robinson (1986) devised a typology of black and Asian settlement for the period 1971–81 and found that the key types of places attracting such settlement in 1971 were the inner areas of London (30.5 per cent), service centres within London (13.9 per cent), traditional industrial conurbations (8.7 per cent), textile towns (7.5 per cent), mixed suburban areas of London (6.8 per cent) and heavy engineering towns (6.3 per cent). When considering changes between 1971 and 1981 in the population born in the New Commonwealth and Pakistan, Robinson noted that the northern textiles towns had gained in relative importance, that inner London had experienced a reduction in its share, that London's suburbs had gained, and that a new category of 'suburbs and suburban centres' had to be added in 1981 to accommodate growth in places such as Milton Keynes, Stockport, Leamington, Basingstoke and Solihull. Robinson (1986) also provided a parallel analysis just for the Asian population.

Table 10.5 uses the same categorisation but relates to the period 1981–86/8 and is based upon Haskey's (1991) data and unpublished 1981 OPCS estimates: it should be noted that the 1981 figures are particularly susceptible to sampling error and therefore that 1981–86/8 comparisons need to be made with caution. What the table suggests is that for the West Indian population the pattern is one of stability. For Indians, there appears to be a reduction in the size of the population in those centres dependent upon manufacturing industry, except for the textile towns which are still recording rapid growth rates. There also seems to be some growth in the suburbs and suburban centres, with the exception of the high status/mixed suburbs category which relies upon particularly anomalous raw numbers. For the Pakistani population, London seems to have lost its attraction, as have conurbations dependent upon traditional industry: this decline has been balanced by sustained growth in the textile and clothing towns.

Finally, it is important to note that the ethnic minority population is not evenly distributed *within* towns and cities and therefore impacts upon the demography of some neighbourhoods or wards much more than others (Robinson, 1989). Both Asians and Afro-Caribbeans tend to be spatially concentrated into the inner-areas of British cities either in modern flatted council housing (West Indians) or in old owner-occupied properties which may have been improved with council grants. Such generalisations do not, of course, apply to all cities or to all individuals, and there is also evidence to show the suburbanisation of Asians into the inner and outer suburbs of certain cities (Werbner, 1979), and the decentralisation of professional West Indians in London. Because decentralisation is occurring in some cities, whilst natural increase is causing greater agglomeration in others, there are no clearly discernible national trends towards segregation or desegregation. Much depends upon local circumstances such as the ethnic minority group concerned, the local housing stock, competition for employment and any extreme

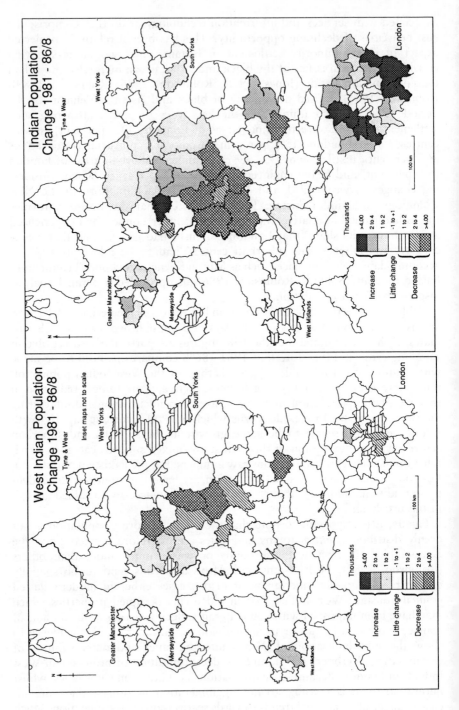

Fig. 10.3 Changes in the spatial distribution of the ethnic minority populations, 1981–86/8 (in thousands).

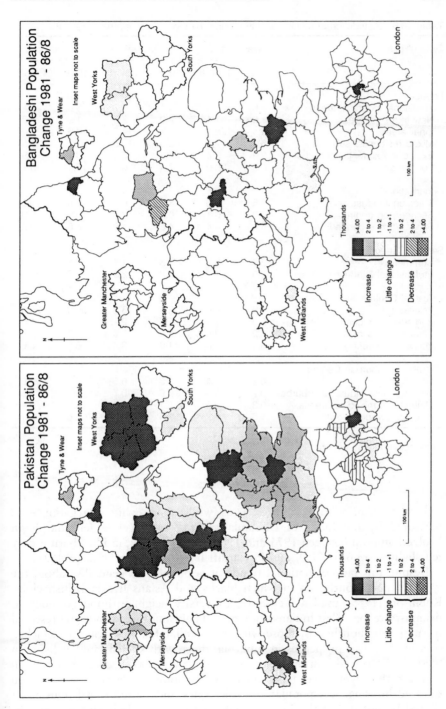

Fig. 10.3   (continued)

Table 10.5 Distribution of ethnic minority groups, by type of place, 1981–1986/8

Type of place	West Indian		Indian		Pakistani	
	1981	1986/8	1981	1986/8	1981	1986/8
London						
Traditional industrial areas	2.6	3.0	4.1	4.7	3.6	4.1
New industrial areas	1.6	1.1	2.9	2.5	1.9	1.3
Inner areas	28.4	28.2	15.7	14.7	7.6	6.4
Mixed residential suburbs	6.3	7.1	13.3	14.5	7.7	4.5
Exclusive residential suburbs	2.9	3.4	5.4	6.2	3.2	1.9
Service centres	20.0	20.0	7.4	6.8	6.4	5.7
Conurbations						
Traditional industrial	7.5	7.6	8.2	6.8	12.5	10.9
Service/port centres	0.3	0.1	0.8	0.3	0.5	1.0
Service/textile centres	5.4	4.4	4.8	4.9	8.6	9.8
Service Centres						
Commercial	2.7	3.2	1.3	1.4	2.1	1.9
Professional & administrative	2.0	1.9	2.8	2.9	2.7	3.8
Industrial Centres						
Textiles	4.9	4.6	8.4	13.3	25.0	29.9
Engineering & commerce	2.3	1.6	2.6	2.7	1.3	1.5
Engineering & railways	4.2	3.6	5.2	5.0	4.7	4.3
Heavy Engineering	6.5	7.0	11.8	8.7	8.2	8.6
Suburbs & Suburban Centres						
New Towns	0.2	0.4	1.0	0.6	0.3	0.6
Working class industrial suburbs	0.7	0.8	0.6	0.7	1.5	1.5
High status mixed residential	0.3	0.3	2.1	0.8	0.2	0.4
Centres of rapid growth	1.3	1.2	1.4	2.3	2.0	2.1

Note: columns may not sum to exactly 100.0 because of rounding.

right wing local political cultures. Thus, whilst Asian-white segregation – as measured by the Index of Dissimilarity at Ward level – fell sharply in Wolverhampton between 1971 and 1981 (51.7 to 42.3), in Blackburn it only fell slightly (53.6 to 51.0) and in Cardiff it actually rose (43.2 to 46.0).

Despite this lack of national trends, it is possible to generalise about the types of neighbourhoods in which blacks and Asians live in British cities. Robinson (1989) used a Newcastle University neighbourhood typology to demonstrate that West Indians, Indians and Pakistanis were under-represented in rural neighbourhoods, those with white-collar families living in detached or semi-detached properties in suburbia, those with skilled families living in 'desirable' terraced housing, and those with low-turnover council houses. In contrast they were over-represented in neighbourhoods with unimproved terraced housing, less popular council estates and flatted social housing. It should be noted that the Indians evidenced the least disadvantaged profile, consonant with the beginning of their recent movement into the middle class (see Robinson, 1988; 1990).

Implications of ethnic minority spatial concentration

There is still a good deal of debate both within theoretical and applied circles about the true significance of the spatial concentration of ethnic minority groups. Boal is perhaps the author who has most successfully bridged the gulf between those who see ethnic concentration as a positive feature which grows naturally out of cultural diversity and which is capable of offering ethnic minorities a range of psychological, social and economic benefits and those who view it as a negative manifestation of white society's racial exclusionism. He argues that segregation can serve a variety of functions and that it can also provide different members of the same community with different benefits at the same time. He suggests that segregation can perform an avoidance function in which a culture is retained and protected, a preservative function in which a culture can be nurtured and can grow as it is passed from generation to generation, a defensive function in which ethnic minorities can use segregation as a way of avoiding verbal abuse or physical harassment, and an attack function in which minority members can use ethnic territory to launch political, economic, or physical attacks upon those in the wider society (Boal, 1978). More recently, Boal (1992) has added a fifth and sixth function to this list. He argues that segregation can also be imposed upon an unwilling minority by an exclusionary majority, or that it allows a temporary spatial stand-off between two groups locked in a struggle to determine which is the dominant or majority group. If Boal is correct, then the true significance of segregation depends upon the uses to which an ethnic minority or the white majority puts it, and the values of the commentator. One needs to bear this in mind when discussing the implications of concentration since different actors have different agenda.

The spatial concentration of ethnic minorities has a wide range of social, political and economic consequences which require more space to discuss than is available here. Ethnic school leavers might be vital to the future of certain industrial sectors and certain local labour markets. Indeed Coleman and Salt (1992) report OPCS statistics which show that 45 per cent of all births in Tower Hamlets were to New Commonwealth or Pakistani mothers in 1987, and a further twelve local administrative units had more than 20 per cent of their births to NCWP mothers. Saggar (1992) has demonstrated the efforts which Ealing Borough Council has made to distribute ethnic pupils more widely between all its schools. The concentration of ethnic minorities in particular localities which are experiencing industrial restructuring and job loss has implications for levels of ethnic unemployment and for those given the responsibility for training. Ethnic minorities might require different local responses from the providers of health care in terms of the facilities on offer, how these facilities may be accessed and whether they are tailored to the needs of those belonging to different cultures. The different demographics of the Pakistani and Bangladeshi populations has implications for the types and size of housing stock which local authorities and, increasingly, housing associations need to make available in those areas in which these groups are concentrated. Local authorities which have within their boundaries ethnic minorities have a statutory obligation under Section 71 of the Race Relations

Act 1976 not only to eliminate racial discrimination in their functioning but also to take on the more positive role of promoting equality of opportunity for all races. Spurred on by the urban disorders of 1980 and 1981 many of the localities of greatest ethnic concentration developed extensive policy agenda and operational structures to combat racial disadvantage (Young and Connelly, 1981), although the agreement of the early 1980s about objectives and methodologies has since given way to conflict, confusion and acrimony (Young, 1990), summarised in Ball and Solomos (1990).

Given that all the issues mentioned above cannot be discussed, two particular themes have been chosen for further treatment, since both have particular salience in 1992. The first concerns the rising political and electoral significance of the ethnic minorities. 1992 saw a General Election, and the question was again asked whether the spatial concentration of the ethnic minorities in particular localities had any impact on the outcome. This issue is crucial to the future of ethnic minorities in this country for, in a democracy, whether a group can influence the outcome of elections determines the extent of that group's influence on the political parties and in turn the willingness of those parties to respond to the group's needs.

In 1988, Indians, Pakistanis, West Indians and Bangladeshis formed perhaps 2.7 per cent of those eligible to vote, thereby suggesting that they would have little influence on any national election. However, some authors argue that the spatial concentration of ethnic voters within a relatively small number of constituencies gives them much greater electoral weight than their numbers alone would suggest. Anwar (1986) is one of the leading champions of this thesis. He argued that, in the mid 1980s, there were 100 Parliamentary constituencies in which ethnic voters formed more than 10 per cent of the total, that in these constituencies ethnic minorities had the power to shape the outcome of any election result and that in the October 1974 General Election they had done exactly that (Anwar and Kohler, 1975). In three constituencies (Brent South, Ealing Southall, and Birmingham Ladywood) ethnic minority voters even formed a numerical majority in 1987 whilst they formed more than a third of the electorate in a further seven constituencies. According to Anwar (1986), the influence of ethnic voters is particularly pronounced in what he terms the marginals, where the number of ethnic voters exceeded the eventual majority of the winning candidate: there were 41 of these in the May 1979 General Election. Exponents of this thesis point to the 1972 parliamentary by-election in Rochdale as an example of the power of ethnic voters: there, Pakistani voters were wooed by Cyril Smith for the Liberals, and they helped him gain an unexpected majority at a time when the Labour party took ethnic votes and also the Rochdale seat for granted (Anwar, 1973). More recently, the Runnymede Trust (1992) has pointed to the substantial swings to sitting Afro-Caribbean MPs in London during the 1992 General Election and argued that these were as significant as the general swing to Labour in the capital. Exponents also point to the growing number of ethnic minority candidates fielded by the major parties (five in 1979, 18 in 1983, 27 in 1987 but only 23 in 1992) and suggest that this is their response to the electoral importace of the ethnic vote: in turn, the number of candidates who have become MPs has also risen from none in 1983 to four in 1987 and six in 1992.

Not all commentators find Anwar's thesis persuasive. Fitzgerald (1986, pp. 226–7) concludes her survey of the 1983 General Election in a very different vein: 'The impact of the "black vote" must realistically be appraised as minimal by comparison with predictions for it .... (It) did not play a significant part in the 1983 General Election and it is, perhaps, time to question whether it ever can realise the potential influence ascribed to it.' She went on to argue that the 'black vote' could only ever become important where three conditions came together, namely where black voters were numerically significant *and* where black voters were organised to act in unison *and* where they could be persuaded to vote along ethnic lines. Fitzgerald did not see such circumstances in the UK in the present or in the near future and thus viewed the black vote as a 'myth'. We might therefore conclude that spatial concentration in particular constituencies might be seen as a necessary but insufficient cause of black and Asian political influence.

The second set of issues which have dominated the early 1990s include Islam, its role in British society, and anti-religious feeling as a new form of racism. As we have noted above, the Pakistani and Bangladeshi populations are the fastest growing of all Britain's main ethnic minority groups, and together with the Arab population of the UK numbered over 600,000 in 1986/8: they are also highly spatially concentrated in particular localities and neighbourhoods. Direct estimates of the number of Muslims in Britain are notoriously difficult to draw up, but both Wahab (1989) and Social Trends (1992) suggest a figure of one million, whilst Peach (1990) thinks three quarters of a million. Three events in the last three years have ensured that this religious minority has received a good deal of prominence and has become the focus of media, political, and research interest. First, in 1989 there was the publication of Salman Rushdie's *Satanic Verses*. The outcry from Muslims in Britain and elsewhere, their very public actions against the book, and the imposition of a *fatwah* by Iranian fundamentalists, all attracted a great deal of media interest in the UK. Perhaps more importantly, the popular and serious media used the event as an opportunity to raise questions about the desirability and likelihood of Britain becoming a truly multiracial society, and the ability of Muslims to participate in this. As *The Daily Telegraph* put it in an editorial on 17 May 1989, 'In the wake of *The Satanic Verses*, there must be increased pessimism about how far different communities in our nation can ever be integrated, or want to be'. The whole event thus emphasised the religious devide and strengthened stereotypes. Scenes of Muslims marching through the streets of some British cities, taking action against bookshops and demanding the death of another citizen created an image for whites of Muslims as militants who were prepared to defy the law to achieve their aims. For Muslims, the Rushdie affair confirmed that Britain along with many other countries had policies and laws which espoused religious tolerance but which seemed to extend these only to Christianity and Judaism, not Islam.

The second event occurred in January 1992 when Dr Kalim Siddiqui inaugurated the first session of the self-styled Muslim Parliament of Great Britain. Although other Asian commentators have been quick to condemn the representatives and constitution of this body (Parekh, 1992), there is no doubt that a national Muslim deliberative body could serve a useful and constructive role in the same way that the Board of Deputies of British Jews and the General Synod of the Church of England 'represent' and voice the interests of their members. Again, however, the tone of the first meeting, as represented by the media, served only to heighten fears that the Muslim community in Britain wished to stand outside mainstream society and was something to be feared. The *Sun* carried a headline 'If you don't like us get out' whilst it also commented that the Parliament was 'a Trojan horse for the spread of the worldwide Islamic revolution'. Even the *Independent* reported that Dr Siddiqui 'talked dramatically of witholding taxes, of civil disobedience, and defying "unjust" laws ...'

The third issue which − seen against the background of those outlined above − seemed to confirm the growing concern about the Muslim presence in Britain was the question of the educational system and how responsive this had been to followers of Islam. Again, it was the Muslim Parliament of Great Britain that crystallised what had been a long standing concern of a number of Britain's ethnic minorities. In March 1992 the parliament published a white paper on the education of Muslims in the UK. It argued that multiculturalism had failed to deliver within the maintained sector, and that most schools resented the 'interference' of Muslims in issues such as collective worship or even in delineating the history syllabus: this was even true where the majority of pupils were Muslims. They suggested that voluntary-aided separate schools were the answer, and that the government was again singling out Muslims for separate and discriminatory treatment. As the paper said 'Jews and Christians who reject secular values can send their children to schools established by a church or a religious foundation and maintained out of public funds. The right is prejudicially denied to Muslim parents ...' (Muslim Parliament of Great Britain, 1992, 00).

The combination of these three issues, the media reporting of them, and the consequent heightening of pre-existing prejudices has made the Muslim population a political concern, both nationally and locally. At a national/supranational level, the Council of Europe in September 1991 adopted recommendation 1162 which states 'Islam has ... suffered and is still suffering from misrepresentation ... through hostile ... stereotypes. Historical errors, educational eclecticism and the over-simplified approach of the media are responsible for this situation. The main consequence ... is that Islam is too often perceived in Europe as incompatible with the principles which are at the basis of modern European society ...' There has also been a growing interest in the spatial distribution within Britain of those who espouse Islam. Violence in Blackburn between Muslim Pakistanis and Hindu Indians in July 1992 added political urgency to the issue. However those concerned with the matter of religious segregation and concentration will have little academic literature to guide them, although Robinson's (1979; 1986) work within cities, and Knott's (1991) work at a national level might both be of value. Equally, local authorities are now becoming acutely aware of the need for accurate local estimates of the size of religious minorities in order that appropriate services can be provided (e.g. halal meals in hospitals). A flurry of field research has resulted.

Conclusion

Quite rightly, since 1979, the government has refused to undertake projection exercises on the size of Britain's ethnic minority population. Given the irresponsible nature of much of the media, the publication of such figures would be inflammatory, would mark a return to the numbers game of the late 1960s and early 1970s and would provide the extreme right with information which could easily be distorted. Even without such projections, however, it is clear that the ethnic minority population in Britain will continue to gain in importance throughout the remainder of this decade and well into the early years of the next century. The contribution of ethnic minorities to local populations will certainly grow, as will their significance in electoral, economic and social terms. Their demographic significance will, as we have argued above, not be general, but will be restricted to particular regions, cities, and parts of cities: indeed ethnic minorities are already *the* key to understanding the demography of certain localities in Britain and this will become true of a greater number of places. Recent work on the spatial redistribution of ethnic minorities through the mechanism of internal migration (Robinson, 1991; 1992b), findings on ethnic social mobility (Robinson, 1988; 1990) and the work on urban typologies described above, also tells us that ethnic minorities will become demographically important in a wider range of types of places than was previously the case. Given all this, the 1991 Census — with the first ethnic question to appear in a British decennial census — is both timely and vitally important to developing our understanding of what is going on within ethnic minorities and how this produces different local outcomes when overlaid onto the unique history, economy, political culture, and society of individual places.

REFERENCES

ACC (1985) *Block Grant Indicators 1986/89*, Association of County Councils, Society of County Treasurers.

Allen, I., Hoog, D. and Peace, S. (1992) *Elderly People: Choice, Participation and Satisfaction*, Policy Studies Institute, London.

Allen, J. and Hamnett, C. (1991) *Housing and Labour Markets*, Unwin Hyman, London.

Anwar, M. (1973) Pakistani participation in the 1972 Rochdale by-election, *New Community*, Vol. 2, pp. 418–23.

Anwar, M. (1979) *The Myth of Return: Pakistanis in Britain*, Heinemann, London.

Anwar, M. (1986) *Race and Politics*, Tavistock, London.

Anwar, M. and Kohler, D. (1975) *Participation of Ethnic Minorities in the General Election, October 1974*, Community Relations Commission, London.

Ashworth, G. and Voogd, H. (1990) *Selling the City*, Belhaven, London.

Association of Directors of Social Services (1992) *Private Residential Care in England and Wales*, ADSS.

Audit Commission (1986a) *Making a Reality of Community Care*, HMSO, London.

Audit Commission (1986b) *Towards Better Management of Secondary Education*, HMSO, London.

Bailey, P. (1982) Falling rolls in the maintained secondary school, *Educational Management and Administration*, Vol. 10, pp. 25–30.

Baldwin, S., Corden, A., Sutcliffe, E. and Wright, K. (1991) *A Survey of Nursing Homes and Hospices*, Centre for Health Economics and Social Policy Research Unit, University of York.

Ball, W. and Solomos, J. (1990) *Race and Local Politics*, Macmillan, Basingstoke.

Benjamin, B. (1968) *Demographic Analysis*, George Allen and Unwin, London.

Bennett, J. (1986) Private nursing homes: contribution to long stay care of the elderly in the Brighton Health District, *British Medical Journal*, Vol. 293, pp. 867–70.

Bennett, R.J. (1980) *The Geography of Public Finance*, Methuen, London.

Bennett, R.J. (1982) *Central Grants to Local Governments: The Political and Economic Impact of the Rate Support Grant in England and Wales*, Cambridge University Press, Cambridge.

Bennett, R.J. (ed.) (1989a) *Decentralisation, Local Governments and Markets: Towards a Post-welfare Agenda*, Clarendon Press, Oxford.

Bennett, R.J. (1989b) Demographic and budgetary influence on the geography of the poll tax: alarm or false alarm? *Transactions of the Institute of British Geographers, New Series*, Vol. 14, No. 3, pp. 400–17.

Bennett, R.J. and Krebs, G. (1988) *Local Business Taxes in Britain and Germany*, Nomos, Baden-Baden.

Bibby, P.R. and Shepherd, J.W. (1991) *Rates of Urbanisation in England 1981–2001* HMSO.

Blakely, E. (1987) Introducing high tech: principles of designing support systems for the formation and attraction of advanced technology firms, *International Journal of Technology Management*, 2.

Boal, F. (1978) Ethnic residential aggregation, in D. Herbert and D. Smith (eds.) *Social Areas in Cities*, Wiley, Chichester.

Boal, F. (1992) Urban ethnic segregation; haven, opportunity or trap?, paper presented at the IGU Population Geography Symposium, Los Angeles.

Bochel, M. (1987) *The Mixed Economy of Residential Care for the Elderly; A Geographical Perspective*, Working Paper No. 2, Age Concern Institute of Gerontology, University of London.

Bolton, N. and Chalkley, B. (1989) Counter-urbanisation: disposing of the myths, *Town and Country Planning*, Vol. 58, pp. 249–50.

Bolton, N. and Chalkley, B. (1990) The rural population turnaround: a case-study of North Devon, *Journal of Rural Studies*, Vol. 6, pp. 29–42.

Bond, J., Gregson, B.A., Atkinson, A. and Newell, D.J. (1989) The implementation of a multi-centred randomised controlled trial in the evaluation of the experimental National Health Service nursing homes, *Age and Ageing*, 18, pp. 96–102.

Bondi, L. (1986) *The Geography and Politics of Contraction in Local Education Provision; a Case Study of Manchester Primary Schools*, Unpublished Ph.D. thesis, University of Manchester.

Bondi, L. (1988) Political participation and school closure: an investigation of bias in local authority decision-making. *Politics and Policy*, Vol. 16, pp. 41–54.

Bosanquet, N. and Grey, A. (1989) *Will You Still Love Me? New Opportunities for Health Services for the Elderly in the 1990s and Beyond*. National Association of Health Authorities, Birmingham.

Bosanquet, N., Laing, W. and Propper, C. (1990) *Elderly Consumers in Britain: Europe's Poor Relations?*, Laing & Buisson Publications, London.

Bracewell, R. (1987) *CACI's Small Area Population, Labour Force and Household Estimates*, Regional Science Association Workshop, 23 October 1987.

Bradford, M.G. (1989) Educational change in the city, in D.T. Herbert and D.M. Smith (eds.) *Social Problems and the City: New Perspectives*, Oxford University Press, Oxford.

Bradford, M.G. (1990) Education, attainment and the geography of choice, *Geography*, Vol. 75, pp. 3–16.

Bradford, M.G. (1991) School performance indicators, the local residential environment and parental choice, *Environment and Planning A*, Vol. 23, pp. 319–32.

Bradford, M.G. and Burdett, F. (1989a) Spatial polarisation of the private education. *Area*, Vol. 21, pp. 47–57.

Bradford, M.G. and Burdett, F. (1989b) Privatisation, education and the North-South divide, in J. Lewis and A. Townsend (eds.) *The North-South Divide: Regional Change in Britain in the 1980s*, Paul Chapman Publishing, London.

Bradford, M.G. and Hall, E. (1992) Privatisation and the blurring of the private-state divide in education: the regional and local consequences of Grant Maintained Schools, *SPA Working paper 18*, Department of Geography, University of Manchester.

Bradshaw, J. and Gibbs, I. (1988) *Public Support for Private Residential Care*, Avebury, Aldershot.

Bramley, G. (1989) The demand for social housing in England in the 1980s. *Housing Studies*, Vol. 4, No. 1, pp. 18–35.

Bramley, G. (1990) *Bridging the Affordability Gap*: report on access to a range of housing options, School for Advanced Urban Studies, University of Bristol.

Bramley, G., le Grand, J. and Low, W. (1989) How far is poll tax a community charge? The implications of service usage evidence, *Policy and Politics*, Vol. 17, pp. 187–206.

Bramley, G. and Paice, D. (1987) *Housing Needs in Non-Metropolitan Areas*, Association of District Councils, London.

Brennan, J. and McGeevor, P. (1987) *Employment or Graduates from Ethnic Minorities*, Commission for Racial Equality, London.

Briault, E. and Smith, F. (1980) *Falling Rolls in Secondary Schools*, NFER, Windsor.

Britton, M. (ed.) (1990) *Mortality and Geography*, OPCS, London.

Brown, C. (1984) *Black and White Britain*, Heinemann, London.

Brown, P. and Ferguson, S.S. (1982) Schools and population change in Liverpool, in W.T.S. Gould and A.G. Hodgkiss (eds.) *The Resources of Merseyside*, Liverpool University Press, Liverpool.

Brown, P., Hirschfield, A. and Batey, P. (1991) Applications of geodemographic methods in the analysis of health conditions incidence data, *Papers in Regional Science*, Vol. 70, pp. 329–44.

Byron, M. (1991) The Caribbean-British Migration Cycle: Migration Goals, Social Networks and Socio-Economic Structure, unpublished D.Phil. thesis in Anthropology and Geography, Oxford University.

Campbell, A., Converse, P. and Rodgers, W. (1976) *The Quality of American Life*, Plenum Press, New York.

Carr-Hill, R. (1990) RAWP is dead – long live RAWP, in A. Culyer, A. Maynard and J. Posnett (eds.) *Competition in Health Care: Reforming the NHS*, Macmillan, London.

Carr-Hill, R. and Sheldon, T. (1991) Designing a deprivation payment for general practitioners: the UPA (8) wonderland, *British Medical Journal*, Vol. 302, pp. 393–6.

Carstairs, V. and Morris, R. (1991) *Deprivation and Health in Scotland*, Aberdeen University Press, Aberdeen.

Castells, M. (1989) *The Informational City*, Blackwell, Oxford.

CBI (1988) *Workforce 2000*, Confederation of British Industry, London.

CBI ERC (1991) Change predicted in structure of mobility, *Relocation News*, No. 20, p. 20.

Chambers, J., Killoran, A., Johnson, K. and Mohan, J. (1990) *Mapping the Epidemic: Coronary Heart Disease in England*, Health Education Authority, London.

Champion, A.G. (1986) Great Britain, in A. Findlay and P. White (eds.) *West European Population Change*, Croom Helm, London.

Champion, A.G. (ed.) (1989a) *Counterurbanisation: The Changing Pace and Nature of Population Deconcentration*, Edward Arnold, London.

Champion, A.G. (1989b) Counterurbanisation in Britain, *Geographical Journal*, Vol. 155, pp. 52–9.

Champion, A.G. (1991) Changes in the spatial distribution of the European population, in *Seminar on Present Demographic Trends and Lifestyles: Proceedings*. Council of Europe, Strasbourg.

Champion, A.G. (1992) Migration in Britain: research challenges and prospects, in A. Champion and A. Fielding (eds.) *Migration Processes and Prospects Vol. 1. Research Progress and Prospects*, Belhaven, London.

Champion, A. and Fielding, A. (eds.) (1992) *Migration Processes and Patterns Vol. 1. Research Progress and Prospects*, Belhaven, London.

Champion, A., Green, A., Owen, D., Ellin, D. and Coombes, M. (1987) *Changing Places: Britain's Demographic, Economic and Social Complexion*, Arnold,

London.

Champion, A. and Stillwell, J. (1991) *Limited Life Working Party on Migration in Britain: Final Report*, Institute of British Geographers, London.

Champion, A.G. and Townsend, A.R. (1990) *Contemporary Britain: A Geographical Perspective*, Arnold, London.

Clark, D. and Cosgrove, J. (1991) Amenities vs labour-market opportunities − choosing the optimal distance to move, *Journal of Regional Science*, Vol. 31, pp. 311−28.

Clarke, S. and Kirby, A. (1990) In search of the corpse: the mysterious case of local politics, *Urban Affairs Quarterly*, Vol. 25, pp. 389−412.

Coleman, D. and Salt, J. (1992) *The British Population*, Oxford University Press, Oxford.

Commission for Racial Equality (1990) *Sorry It's Gone*, Commission for Racial Equality, London.

Compton, P. (1990) Review of Joshi, H. (ed.) *The Changing Population of Britain*, *Progress in Human Geography*, Vol. 14, pp. 462−4.

Congdon, P. and Batey, P. (eds.) (1989) *Advances in Regional Demography*, Belhaven Press, London.

Corden, A. (1988) *Supplementary Benefit Claimants in Residential Care and Nursing Homes in North Yorkshire and Somerset: Changes in Characteristics 1985/6−88*, Working Paper DHSS 494, Social Policy Research Unit, University of York.

Corden, A. (1989a) *Background Data on North Yorkshire and Somerset*, Working Paper DHSS 401, Social Policy Research Unit, University of York.

Corden, A. (1989b) *Developments in the Supply of Places for Long Term Care in the Independent Sector: Findings from a Monitoring Study in North Yorkshire and Somerset*, Working Paper DHSS 515, Social Policy Research Unit, University of York.

Corden, A. (1992) Geographical development of the long-term care market for elderly people, *Transactions of the Institute of British Geographers New Series*, Vol. 17, No. 1, pp. 80−94.

Corlett, S., Collinson, J. and Pleace, N. (1991) *Social Security and Residential Nursing Home Care*, Northumberland Social Services Department, County Hall, Morpeth.

Corner, I.E. (1991) *Household Demography and the Effective Demand for New Housing*, Paper presented to the European Symposium on Management, Quality and Economics in Housing and Other Building Sectors, Lisbon 30 September−4 October 1991.

Cumberlege, J. (1986) *Neighbourhood Nursing: A Focus for Care*, HMSO, London.

Curtis, S. (1990) Use of survey data and small area statistics to assess the link between individual morbidity and neighbourhood deprivation, *Journal of Epidemiology and Community Health*, Vol. 44, pp. 62−8.

Curtis, S. and Taket, A. (1989) Locality planning for health care: a case study in east London, *Area*, Vol. 21, pp. 357−64.

Curtis, S. and Woods, K. (1984) Health care in London: planning issues and the contribution of local morbidity surveys, in M. Clarke, (ed.) *Planning and Analysis in Health Care Systems*, Pion, London.

Darton, R., Sutcliffe, E. and Wright, K.G. (1992) *Private and Voluntary Residential and Nursing Homes: A Report of a Survey by PSSRU/CHE, PSSRU*, University of Kent at Canterbury (forthcoming).

Davies, B. (1968) *Social Needs and Resources in Local Service Administration*, Michael Joseph, London.

Davies, B.P., Barton, A.J., McMillan, I.S. (1974) *Variations in Children's Services Among British Urban Authorities*, Bell, London.

Davies, B.P., Barton, A.J., McMillan, I.S. and Williamson, V.K. (1971) *Variations in Services for the Aged*, Bell, London.

Davies, C. (1987) Things to come: the NHS in the next decade, *Sociology of Health and Illness*, Vol. 9, pp. 309–27.

Day, R. and Walmsley, D. (1981) Residential preferences in Sydney inner suburbs: a study in diversity, *Applied Geography*, Vol. 1, pp. 185–97.

Department for Education (1992) *Choice and diversity: a new framework*, HMSO, London.

Department of Education and Science (1977) *Falling Numbers and School Closures*, Circular 5/77, HMSO, London.

Department of Education and Science (1978) *School Population in the 1980s*, Report on Education 92, HMSO, London.

Department of Education and Science (1979) *Trends in School Population*, Report on Education 96, HMSO, London.

Department of Employment (1988) *Employment for the 1990s*, Cmnd. 540, HMSO, London.

Department of Employment (1991) Labour force trends: the next decade, *Employment Gazette*, Vol. 99, pp. 269–80.

Department of the Environment (1980a) *Development Control – Policy and Practice*, Circular 22/80, HMSO, London.

Department of the Environment (1980b) *Land for Private Housebuilding*, Circular 9/80, HMSO, London.

Department of the Environment (1992a) *Planning Policy Guidance No. 1 General Policy and Principles*, HMSO, London.

Department of the Environment (1992b) *Planning Policy Guidance No. 3 Housing*, HMSO, London.

Department of the Environment (1992c) *Planning Policy Guidance No. 12 Development Plans and Regional Guidance*, HMSO, London.

Department of Health (1991) *Health and Personal Social Services Statistics for England*, HMSO, London.

DHSS (1985) *Report of the Steering Group on Health Service Information (the Korner Committee)*, DHSS, London.

DHSS (1988) *Review of the Resource Allocation Working Party Formula: Report of the NHS Management Board*, DHSS, London.

Dicks, M.J. (1988) *The Demographics of Housing Demand: Household Formation and the Growth of Owner-Occupation*, Bank of England Discussion Paper No. 32.

Edwards, T., Fitz, J. and Whitty, G. (1989) *The State and Private Education: an Evaluation of the Assisted Places Scheme*, Falmer, London.

Elias, D.P.B. and Owen, D.W. (1989) *People and Skills in Coventry: An Audit of the Education, Training, Qualifications and Work Experience of Coventry People*, Coventry City Council, Coventry.

Employment Department Group (1992) What are employers doing to defuse the 'demographic time bomb'?, Skills and Enterprise Briefing, Issue 21/92, *Skills and Enterprise Network*, Nottingham.

Employment Department Yorkshire and Humberside (1991) *Planning for a Changing Labour Market: Labour Market Assessment 1991/92*, Employment Department Yorkshire and Humberside Regional Office, Leeds.

Ermisch, J. (1983) *The Political Economy of Demographic Change*, Heinemann, London.

Ermisch, J. (1990) *Fewer Babies, Longer Lives*, Joseph Rowntree Foundation,

York.

Ermisch, J. (1991) An ageing population, household formation and housing *Housing Studies*, Vol. 6, No. 4, pp. 230—9.

European Commission (1990) *Green Paper on the Urban Environment*, COM(90)218 EC.

Fielding, A.J. (1989) Inter-regional migration and social change, *Transactions, Institute of British Geographers New Series*, Vol. 14, pp. 24—36.

Fielding, A.J. (1992a) Migration and culture, in A.G. Champion and A.J. Fielding (eds.) *Migration Processes and Patterns, Vol. 1: Research Progress and Prospects*, Belhaven, London.

Fielding, A.J. (1992b) Migration and social change, in J. Stillwell, P. Rees, and P. Boden (eds.) *Migration Processes and Patterns Vol. 2 Population Redistribution in the United Kingdom*, Belhaven, London.

Findlay, A. and Rogerson, R. (1991) Voting with their feet? Migration and quality of life in Britain in the 1980s, *APRU Discussion Paper 91/4*.

Findlay, A., Rogerson, R. and Morris, A. (1988) Where to live in Britain in 1988, *Cities*, 5, pp. 268—76.

Fitzgerald, M. (1986) The parties and the black vote, in I. Crewe and M. Harrop (eds.) *The General Election Campaign of 1983*, Cambridge University Press, Cambridge.

Fryer, P. (1984) *Staying Power: The history of black people in Britain*, Pluto Press, London.

Fuguitt, G. (1991) Commuting and the rural-urban hierarchy, *Journal of Rural Studies*, Vol. 7, pp. 459—66.

Fyson, A. (1992) Time for review, *The Planner*, Vol. 78, 6, p. 3.

Gabbay, J. and Stevens, A. (1991) Needs assessment, needs assessment *Health Trends*, Vol. 23, pp. 20—3.

Gatrell, A., Dunn, C. and Boyle, P. (1991) The relative utility of the Central Postcode Directory and Pinpoint Address Code in Geographical Information Systems, *Environment and Planning A*, Vol. 23, pp. 1447—58.

Gatrell, A. and Naumann, I. (1992) Hospital location planning: a pilot GIS study, forthcoming, in *Proceedings, Mapping Awareness 1992*.

Gibson, J.G. (1990) *The Politics and Economics of the Poll Tax: Mrs. Thatcher's Downfall*, EMAS Ltd., Warley.

Gilje, E. (1974) Which projection to choose? A user's dilemma. Paper presented at PTRC Summer Annual Meeting.

Gillespie, A.E. (1987) Telecommunications and peripheral regions, Paper presented to British Association Annual Conference, Belfast, 27th August.

Goddard, J. (1991) New technology and the geography of the UK information economy, in J. Brotchie *et al.* (eds.) *Cities of the 21st Century*, Longman Cheshire, Harlow, pp. 191—213.

Goodchild, B. (1992) Land allocation for housing, *Housing Studies*, Vol. 7, 1, pp. 45—55.

Gordon, I.R. (1992) Modelling approaches to migration and the labour market, in A.G. Champion and A.J. Fielding (eds.) *Migration Processes and Patterns Vol. 1. Research Progress and Prospects*, Belhaven, London.

Gould, M. (1992) The use of GIS and CAC by health authorities: results from a postal questionnaire, *Area*, 24, (4) pp. 391—401.

Gould, W.T.S. and Lawton, R. (1986) *Planning for Population Change*, Croom Helm, London.

Green, A.E. and Hasluck, C. (1991) *Spatial Aspects of Skill Shortages: Analysis of the Skills Monitoring Survey*, Institute for Employment Research, University of Warwick, Coventry.

Green, A.E. and Owen, D.W. (1990a) The development of a classification of travel-to-work areas, *Progress in Planning*, Vol. 34, pp. 1−91.

Green, A.E. and Owen, D.W. (1990b) Long-term unemployment: JUVOS analysis, *Employment Department Research Paper*, No. 72, Department of Employment, London.

Green, A.E. and Owen, D.W. (1991) Local labour supply and demand interactions in Britain during the 1980s, *Regional Studies*, Vol. 25, pp. 295−314.

Green, A.E., Owen, D.W., Champion, A.G., Goddard, J.B and Coombes, M.G. (1986) What contribution can labour migration make to reducing unemployment? in P.E. Hart (ed.) *Unemployment and Labour Market Policies*, Joint Studies in Public Policy 12, Gower Press, Aldershot.

Greenwood, M. and Hunt, G. (1989) Jobs versus amenities in the analysis of metropolitan migration *Journal of Urban Economics*, 25, pp. 1−16.

Grundy, E. (1987) Retirement migration and its consequences in England and Wales, *Ageing and Society*, Vol. 7, pp. 57−82.

Grundy, E. (1989) Longitudinal perspectives on the living arrangements of the elderly, in M. Jefferys (ed.) *Growing Old in the Twentieth Century*, Routledge, London.

Hale, D. (1991) The healthcare industry and GIS, *Mapping Awareness*, Vol. 5, pp. 36−9.

Hamnett, C. and Mullings, B. (1992) The distribution of public and private residential homes for elderly persons in England and Wales, *Area*, Vol. 24, No. 2, pp. 130−44.

Hansard (1990) 19 December, col. 172.

Hansard (1991) 11 November, col. 402.

Hardill, I. and Green, A.E. (1991) Women returners: and view from Newcastle, *Employment Gazette*, Vol. 99, pp. 147−52.

Harrogate Health Authority (1987) *Nursing Homes and Their Residents*, Harrogate Health Authority and Harrogate Community Health Council.

Harrop, A. and Grundy, E.M.D. (1991) Geographical variations in moves into institutions among the elderly in England and Wales, *Urban Studies*, Vol. 28, No. 1, pp. 65−86.

Haskey, J. (1991) Ethnic minority populations resident in private households − estimate by county and district, *Population Trends*, Vol. 63, pp. 22−36.

Haskey, J. and Kiernan, K. (1989) Cohabitation in Great Britain. *Population Trends*, No. 58, pp. 23−32.

Haughton, G. (1990) Skills shortage and the demographic time-bomb: labour market segmentation and the geography of labour, *Area*, Vol. 22, pp. 339−45.

HC (1991a) *The Financing of Private Residential and Nursing Home Fees*, Social Security Committee, Minutes of Evidence 7 May 1991, para 44, HMSO, London.

HC (1991b) *The Financing of Private Residential and Nursing Home Fees*, Fourth Report of Social Security Committee, para 47, HMSO, London.

Healy, G. and Kraithman, D. (1989) *Women Returners in the North Hertfordshire Labour Market*, Local Economy Research Unit, Hatfield Polytechnic, Hertford.

Healey, M.J. (ed.) (1991) *Economic Activity and Land Use: The Changing Information Base for Local and Regional Studies*, Longman, Harlow.

Hepworth, M., Green, A. and Gillespie, A. (1987) The spatial division of information labour in Great Britain *Environmental and Planning A*, Vol. 19, pp. 793−806.

HM Treasury (1984) *The Next Ten Years: Public Expenditure and Taxation into the 1990s*, HMSO, London, Cmnd. 9189.

Hobcraft, J. and Joshi, H. (1989) Population matters, in H. Joshi (ed.) *The Changing Population of Britain*, Blackwell, Oxford.

Hoffmann-Nowotny, H.-J. and Fux, B. (1991) Present demographic trends in Europe, in *Seminar on Present Demographic Trends and Lifestyles: Proceedings*. Council of Europe, Strasbourg.

Hogarth, T. and Barth, M.C. (1991) *Age Works: Why Employing the Over 50s Makes Good Business Sense*, Institute for Employment Research, University of Warwick, Coventry.

Hopflinger, F. (1991) The future of household and family structures in Europe, *Seminar on Present Demographic Trends and Lifestyles: Proceedings*. Council of Europe, Strasbourg.

Housebuilders Federation (HBF) (1988) *Homes, Jobs and Land: the Eternal Triangle*, Housebuilders Federation, London.

Housing Services Advisory Group (HSAG) (1977) *The Assessment of Housing Requirements*, DoE: London.

Hudson, R. and Williams, A.M. (1989) *Divided Britain*, Belhaven Press, London.

Hunter, D.J., McKeganey, N.P. and Macpherson, I.A. (1988) *Care of the Elderly: Policy and Practice*, Aberdeen University Press.

IFF (1992) *Commuting Behaviour*, Market Research Report prepared for TEED South East, IFF, London.

Institute for Employment Research (1991) *Review of the Economy and Employment*, Institute for Employment Research, University of Warwick, Coventry.

Institute of Housing (1992) *A Radical Consensus: New Ideas for Housing in the 1990s*, Institute of Housing.

Jarman, B. (1983) Identification of Underprivileged Areas, *British Medical Journal*, Vol. 286, pp. 1750−9.

Jones, H. (1990) *Population Geography*, Paul Chapman, London.

Jones, K. and Moon, G. (1987) *Health, Disease and Society: an Introduction to Medical Geography*, Routledge, London.

Jones, K. and Moon, G. (1990) A multilevel approach to immunisation uptake, *Area*, Vol. 22, pp. 264−71.

Jones, P. (1978) The distribution and diffusion of the coloured population in England and Wales, 1961−71, *Transactions, Institute of British Geographers*, Vol. 3, pp. 515−33.

Joshi, H. (ed.) (1989) *The Changing Population of Britain*, Blackwell, Oxford.

Joshi, H. and Diamond, I. (1990) Demographic projections: who needs to know? in *Population Projections: Trends, Methods and Uses*, Occasional Paper 38, OPCS, London.

Kaa, D.J. van de (1987) Europe's second demographic transition, *Population Bulletin*, 42, 1, pp. 1−57.

Keeble, D.E. (1989) The dynamics of European industrial counterurbanisation in the 1980s, *Geographical Journal*, Vol. 155, pp. 70−4.

Keeble, D.E. (1990) Small firms, new firms and uneven regional development in the United Kingdom, *Area*, Vol. 22, pp. 234−45.

Keeble, D.E. and Gould, P. (1985) Entrepreneurship and manufacturing firm formation in rural regions, in M. Healey and B. Ilbery (eds.) *The Industrialisation of the Countryside*, Geobooks, Norwich, pp. 197−219.

King, A. (1990) I fight therefore I am, in *The world in 1991, Economist Publications*, London, pp. 27−8.

King, D. (1991) The 'demographic bulldozer' − myth or reality? The Housebuilders Federation, London.

King, D.M. (1989) Accountability and equity in British local finance: the poll tax, in R.J. Bennett (ed.) *Decentralisation, Local Governments and Markets: Towards a Post Welfare Agenda*, Clarendon Press, Oxford.

Kirby, A. (1990) On social representations of risk, in A. Kirby (ed.) *Nothing to Fear*, Arizona University Press, Tucson.

Kivell, P., Turton, B. and Dawson, B. (1990) Neighbourhoods for health service administration *Social Science and Medicine*, Vol. 30, pp. 701–11.

Knight, D. (1992) *Making Use of the 1991 Census: An Area Classification for the NHS*, Mersey RHA, Liverpool.

Knott, K. (1991) Bound to change? The religions of South Asians in Britain, in S. Vertovec (ed.) *Aspects of the South Asian Diaspora*, Oxford University Press, Delhi.

Laing and Buisson (1991) *Care of Elderly People*, Market Survey. Fifth Edition 1991–92, Laing & Buisson Publications, London.

Larder, D., Day, P. and Klein, R. (1986) *Institutional Care for the Elderly: The Geographical Distribution of the Public/Private Mix in England*, Bath Social Policy Paper No. 10, Centre for the Analysis of Social Policy, University of Bath.

Lawton, R. (1986) Planning for people, in W.T.S. Gould and R. Lawton (eds.) *Planning for Population Change*, Croom Helm, London.

Lelievre, E. (1991) Household and family changes at the beginning of the 1980s in Great Britain *APRU Discussion Paper* 91/2, University of Glasgow, Glasgow.

Lomas, G. (1973) *Census 1971: The Coloured Population of Great Britain*, Runnymede Trust, London.

London Research Centre (1988) *Access to Housing in London: A Report Based on the Results of the London Housing Survey 1986–7*, LRC, London.

Mabogunje, A.L. (1970) Systems approach to a theory of urban-rural migration *Geographical Analysis*, Vol. 2, pp. 1–18.

McLoughlin, J. (1990) *The Demographic Revolution*, Faber & Faber, London.

Maguire, D. (1991) An overview and definition of GIS, in D. Maguire, M. Goodchild and D. Rhind (eds.) *Geographical Information Systems: Volume 1: Principles*, Longman, Harlow.

Marsh, C., Arber, S., Wrigley, N., Rhind, D. and Bulmer, M. (1988) The views of academic social scientists on the 1991 UK Census of Population: a report of the Economic and Social Research Council working group, *Environment and Planning A*, Vol. 20, pp. 851–89.

Massey, D. (1984) *Spatial Divisions of Labour*, Macmillan, London.

Mays, N. and Bevan, G. (1987) *Resource Allocation in the NHS: A Review of the Methods of the Resource Allocation Working Party*, Bedford Square Press, London.

Meadows, P., Cooper, H. and Bartholomew, R. (1988) *The London Labour Market*, HMSO, London.

Metcalf, H. and Leighton, P. (1989) The under-utilisation of women in the labour market, *IMS Report*, No. 172, IMS, University of Sussex.

Mohan, J. and Maguire, D. (1985) Harnessing a breakthrough to meet the needs of healthcare, *Health Services Journal*, 9 May, pp. 580–1.

Moon, G. and Twigg, L. (1988) Health education and baseline data: issues and strategies in nutrition campaigning, *Social Science and Medicine*, 26, pp. 173–8.

Moulden, M. and Bradford, M. (1984) Influences on educational attainment: the importance of the local residential environment, *Environment and Planning A*, Vol. 16, pp. 49–66.

Murphy, M. (1989) Housing the people; from shortage to surplus? in H. Joshi, (ed.) *The Changing Population of Britain*, Blackwell, Oxford.

Murphy, M. and Hobcraft, J. (eds.) (1991) *Population Research in Britain*, Supplement to Population Studies Volume 45, London: Population Investigation Committee.

Murray, N. (1985) Turning the tide, *Community Care*, No. 588, pp. 23–5.

Muslim Parliament of Great Britain (1992) *White Paper on Muslim Education in Great Britain*, Muslim Parliament of Great Britain, London.

National Association of Health Authorities and Trusts (NAHAT) (1992) *Briefing Paper on Resource Allocation*, NAHAT, Birmingham.

National Economic Development Office/Training Agency (1989) *Defusing the Demographic Time Bomb*, NEDO, London.

National Economic Development Office/Training Commission (1988) *Young People and the Labour Market: A Challenge for the 1990s*, NEDO, London.

Niner, P. (1989) *Housing Needs in the 1990s*, Housing Forum.

OECD (1988) *Ageing Populations: The Social Policy Implications*, Organisation for Economic Cooperation and Development, Paris.

Offord, J. (1987) *The Fiscal Implications of Differential Population Change for Local Authorities in England and Wales 1971–1981*, unpublished Ph.D. thesis, University of Cambridge.

OPCS (1990) *Monitor PP1 90/1*, OPCS, London.

Openshaw, S., Charlton, M., Craft, A. and Birch, J. (1988) Investigation of leukaemia clusters by use of a Geographical Analysis Machine, *Lancet*, i, pp. 272–3, 6th February.

Owen, D.W. and Green, A.E. (1992) Labour market experience and occupational change amongst ethnic groups in Great Britain, *New Community*, Vol. 19, No. 1, pp. 7–29.

Panayi, P. (1992) Refugees in Twentieth-century Britain: A brief history, in V. Robinson (ed.) *The International Refugee Crisis: British and Canadian Responses*, Macmillan, Basingstoke.

Parekh, B. (1992) A Muslim deliberative body, *Runnymede Bulletin*, Vol. 253, p. 4.

Parr, J. (1987) Interaction in an urban system: aspects of trade and commuting, *Economic Geography*, 63, pp. 223–40.

Patterson, S. (1963) *Dark Strangers: A Study of West Indians in London*, Tavistock, London.

Peach, C. (1990) The Muslim population of Great Britain, *Ethnic and Racial Studies*, Vol. 13, pp. 414–9.

Pearson, M., Smith S. and White, S. (1989) Demographic influences on public spending, *Fiscal Studies*, Vol. 10, No. 2, pp. 48–65.

Phillimore, P. (1990) Mortality variations within two poor areas in north east England, *The Statistician*, Vol. 39, pp. 373–83.

Phillips, D. (1986) *What Price Equality?*, Greater London Council, London.

Phillips, D.R. and Vincent, J. (1988) Privatising residential care for elderly people: The geography of developments in Devon, *Social Science and Medicine*, Vol. 26, pp. 37–47.

Phillips, J. and Davies, M. (1990) *Admission to Residential Care*, Norwich Social Work Monographs, University of East Anglia.

Pooley, C. (1977) The residential segregation of migrant communities in mid-Victorian Liverpool, *Transactions, Institute of British Geographers*, Vol. 2, pp. 364–83.

Quinn, D.J. (1986) Accessibility and job search: a study of unemployed school leavers, *Regional Studies*, Vol. 20, pp. 163–74.

Raab, G. and Adler, M. (1988) A tale of two cities: the impact of parental choice on admissions to primary schools in Edinburgh and Dundee, in L. Bondi and M.H. Matthews *Education and Society: Studies in the Politics, Sociology and Geography of Education*, Routledge, London.

Rees, P., Stillwell, J. and Boden, P. (1992) Internal migration in the 1980s in J. Stillwell, P. Rees and P. Boden (eds.) *Migration Processes and Patterns Vol. 2 Population Redistribution in the United Kingdom*, Belhaven, London, pp. 1–10.

Rex, J. (1970) *Race Relations in Sociological Theory*, Weidenfeld and Nicholson, London.

Robinson, V. (1979) *The Segregation of Immigrants in a British City*, Oxford University School of Geography, Oxford.

Robinson, V. (1986) *Transients, Settlers and Refugees: Asians in Britain*, Clarendon Press, Oxford.

Robinson, V. (1987) Spatial variability in attitudes towards race in the UK, in P. Jackson (ed.) *Race and Racism*, Allen & Unwin, London.

Robinson, V. (1988) The new Indian middle class in Britain, *Ethnic and Racial Studies*, Vol. 11, pp. 456–73.

Robinson, V. (1989) Economic restructuring, the urban crisis and Britain's black population, in D. Herbert and D. Smith (eds.) *Social Problems and the City*, (2nd. ed.) Oxford University Press, Oxford.

Robinson, V. (1990) Roots to mobility: the social mobility of Britain's black population, *Ethnic and Racial Studies*, Vol. 13, pp. 274–86.

Robinson, V. (1991) Good-bye yellow brick road: the spatial mobility and immobility of Britain's black and Asian population, *New Community*, Vol. 17, pp. 313–31.

Robinson, V. (1992a) The invisible minority: Chinese in the UK, *Revue Europeenne des Migrations Internationales*, Vol. 8, in press.

Robinson, V. (1992b) Move on up: the mobility of Britain's Afro-Caribbean and Asian populations, in J. Stillwell, P. Rees and P. Boden (eds.) *Migration Processes and Patterns Vol. 2*, Belhaven, London.

Rogerson, R. (1989) Measuring quality of life: methodological issues and problems *APRU Discussion Paper 89/2*, University of Glasgow, Glasgow.

Rogerson, R., Findlay, A., Coombes, M. and Morris, A. (1989) Indicators of quality of life: some methodological issues, *Environment and Planning A*, Vol. 21, pp. 1655–66.

Rogerson, R., Findlay, A., Morris, A. and Paddison, R. (1990) *Quality of Life in Britain's District Councils*, Glasgow Quality of Life Group, Glasgow.

Rudzitis, G. (1991) Migration, sense of place and non-metropolitan vitality *Urban Geography*, 12, pp. 80–8.

Runneymede Trust (1992) Ethnic minority candidates in the 1992 General Election, *Runneymede Bulletin*, Vol. 255, pp. 5–7.

Rydin, Y. (1986) *Housing Land Policy*, Gower, London.

Saggar, S. (1992) *Race and Public Policy*, Avebury, London.

Salt, J. (1990) Organisational labour migration, in J. Johnson and J. Salt (eds.) *Labour Migration*, Fulton, London.

Senior, M. (1991) Deprivation payments to GPs: not what the doctor ordered *Environment and Planning C*, Vol. 9, pp. 79–94.

Senior, M. (1992) Deprivation payments to GPs: improving the Jarman index for use with 1991 Census data. Paper presented to the Annual Conference of the Institute of British Geographers, Swansea, January 1992.

Shaw, C. (1988) Components of growth in the ethnic minority population, *Population Trends*, Vol. 52, pp. 26–30.

Smith, A. (1989) Gentrification and the spatial constitution of the State: the restructuring of London's Docklands, *Antipode*, Vol. 21, pp. 232–60.

Smith, C. (1986) Residential accomodation for the elderly in the Nottingham Health Authority, *East Midlands Geographer*, Vol. 9, pp. 45–52.

Smith, D.M. (1976) *Human Geography: A Welfare Approach*, Edward Arnold, London.

Smith, E. (1990) *Skill Needs in Britain*, IFF, London.

Smith, N. and Williams, P. (1986) (eds.) *Gentrification and the City*, Allen & Unwin, London.

Social Trends (1992) Vol. 22, HMSO, London.

Somerset County Council (1989) *Elderly People in Residential Care: A Study of Dependency Among Residents in Residential Care Homes in the Private and Statutory Sectors, Specialised Sheltered Housing, Nursing Homes and Long Stay Hospital Provision*, Social Services Department, Taunton.

Spence, A. (1990) Labour force outlook to 2001, *Employment Gazette*, Vol. 98, pp. 186–98.

Stillwell, J. and Boden, P. (1989) Internal migration: the United Kingdom, in J. Stillwell and H. Scholten (eds.) *Contemporary Research in Population Geography*, Kluwer, Dordrecht.

Stillwell, J., Boden, P. and Rees, P. (1990) Trends in internal net migration in the UK: 1975–1986, *Area*, Vol. 22, pp. 57–65.

Stillwell, J., Rees, P. and Boden, P. (eds.) (1992) *Migration Processes and Patterns Vol. 2. Population Redistribution in the United Kingdom*, Belhaven, London.

Sullivan, O. (1986) Housing movements of the divorced and separated, *Housing Studies*, 1, pp. 35–48.

Thompson, E.J. (1974) Population projections for metropolitan areas, *GLC Intelligence Quarterly*, No. 28, September.

Todd, J.E. (1990) *Care in Private Homes*, OPCS, HMSO, London.

Townsend, P. and Davidson, N. (1982) *Inequalities in Health: The Black Report*, Penguin, London.

Townsend, P., Phillimore, P. and Beattie, A. (1988) *Deprivation and Ill Health: Inequality and the North*, Croom Helm, London.

Training Agency (1990) *Training in Britain*, HMSO, London.

Visram, R. (1986) *Ayahs, Lascars and Princes: Indians in Britain 1700–1947*, Pluto Press, London.

Wade, B., Sawyer, L. and Bell, J. (1983) *Dependency with Dignity*, Occasional Papers in Social Administration No. 68, Bedford Square Press/NCVO, London.

Wahab, I. (1989) *Muslims in Britain – Profile of a Community*, Runneymede Trust, London.

Waite, G. and Pike, G. (1989) School leaver decline and effective local solutions, *IMS Report*, No. 178, IMS, University of Sussex.

Walford, G. (1991) City Technology Colleges: a private magnetism, in Walford, G. (ed.) *Private Schooling: Tradition, Change and Diversity*, Paul Chapman Publishing, London.

Warnes, A. (1990) Geographical questions in gerontology: needed directions for research, *Progress in Human Geography*, Vol. 14, pp. 24–56.

Warnes, A.M. (1992) Age-related variation and temporal change in elderly migration, in A. Rogers (ed.) *Elderly Migration and Population Redistribution: A Comparative Study*, Belhaven, London.

Warnes, A. and Law, C.M. (1984) The elderly population of Great Britain, *Transactions, Institute of British Geographers New Series*, Vol. 9, pp. 37–59.

Waterman, S. and Kosmin, B. (1986) The Jews of London, *Geographical Magazine*, January, pp. 21–7.

Welsh Office (1991) *Health and Personal Social Service Statistics for Wales*, Cardiff, Welsh Office.

Werbner, P. (1979) Avoiding the ghetto: Pakistani migrants and settlement shifts in Manchester, *New Community*, Vol. 7, pp. 376–89.

Westland, P. (1992) Can't pay, won't pay, don't care . . ., *Health Service Journal*, 20 February, p. 24.

White, M. (1991) *Against Unemployment*, PSI, London.

Whitehead, M. (1987) *The Health Divide*, Health Education Council, London.

Wilcox, S. (1990) *The Need for Social Rented Housing in England in the 1990s*, Institute of Housing, London.

Williams, A. and Jobse, P. (1990) Economic and quality of life considerations in urban-rural migration, *Journal of Rural Studies*, Vol. 6, pp. 187–94.

Woodhead, K.R. (1985) Population Projections, in K.I. Hudson, R.J. Masters, K.S. Powell and J.D. Shortridge (eds.) *Information Systems for Policy Planning in Local Government*, Harlow, Longman.

Woodhead, K. and Dugmore, K. (1990) Local and Small Area Projections, in *Populations Projections: Trends, Methods and Uses*, Occasional Paper 38, OPCS, London.

Wright, K. (1984) *Contractual Arrangements for Geriatric Care in Nursing Homes*, Discussion Paper No. 4, Centre for Health Economics, University of York.

Young, K. (1990) Approaches to policy development in the field of Equal Opportunities, in W. Ball and J. Solomos (eds.) *Race and Local Politics*, Macmillan, Basingstoke.

Young, K. and Connelly, N. (1981) *Policy and Practice in the Multi-racial City*, Policy Studies Institute, London.

Zelinsky, W. (1971) The hypothesis of the mobility transition, *Geographical Review*, 61, pp. 219–49.

Zukin, S. (1992) The city as a landscape of power, in L. Budd and S. Whimster (eds.) *Global Finance and Urban Living*, Routledge, London.

INDEX

ACORN 140, 144
activity rate effect 84, 87
activity rates 96, 97, 99
affordability 112
Afro-Caribbeans, see ethnic minority
    populations
age structure 3, 12–13, 22–4, 29,
    30–31, 62, 85–6
alternative methods 26
Asians, see ethnic minority populations
Assisted Places Scheme 80

baby boom and bust 2, 18–19, 57
baby bust cohorts 105
Bangladeshis, see ethnic minority
    populations
banking 28
Bedfordshire 15
Berkshire 67
Birmingham 96, 99
birth rate 2–3, 6, 101
    effect on school rolls 65–70
births 165
Black Report 144
Bradford 34, 97
Brighton 125, 131
British Household Panel Study 21, 48
British Rail 154
'brownfield' sites 115
Brixton 153
Buckinghamshire 9, 13

CACI 28
Cambridgeshire 9
capital finance for local government 52

care homes 119–35
    choice of homes 129–31, 133–5
    geographical distribution 121–2
    growth in numbers 121
    moves between homes 131–3
    moves into homes 124–9
    numbers of elderly in 120–4
catchment areas for schools 66, 73, 74,
    76
Census 1981 124
Census 1991 2, 5, 7, 20, 21, 31–2, 48,
    100, 121, 125, 137, 141, 143, 146,
    147
Census 27
central government grants 52, 59
Central Postcode Directory 145
Chelmer Population and Housing Model
    106
Chinese 152, 154
    see also ethnic minority populations
Citizens' Charter 148
City Technical Colleges 76, 80
Cleveland 82, 142
Clydebank 11
Clwyd 128
community care planning 119, 133–5,
    148
commuting 38, 95
conurbations 37
Cornwall 9
council house sales 111
Council of Europe 168
Council Tax 53
counterurbanisation 4, 117
Coventry 97

data sources 27
demographic bulldozer 101–5
demographic time-bomb 4, 19, 102
  and the labour market 83–100
  national perspective 84–6
  regional trends 86–9
deprivation indices 139, 141–3
Development Plan system 114–5
Devon 128
'difficult to let' estates 113
Direct Grant Schools 80
divorces 3
Docklands 28
Dorset 9
Durham 82

early retirement 97
East Anglia 8, see also regional trends
East Sussex 13, 81
economic activity rate 84
education 18, 63, 64–82
  opting out 73–9
  private/state continuum 79–82
  see also school rolls
Education Act, 1980, 73
Education, 1992 White Paper 79
Education Reform Act 1988 18, 64,
  73–6
elections 166
electoral roll 27
elderly 12–13, 19, 119–135, 141
  choice in move into care 129–31
  factors affecting choice of care home
    133–5
  in longterm care 120–3
  moves into institutions 124–9
  moving between care homes 131–3
elderly care 2
Employment for the 1990s, 1988 White
  Paper 95
Enumeration Districts 29
ethnic minorities populations 2, 13,
  14–18, 20, 78, 94, 96–7, 99, 141
  150–69
  components of change 152
  data 151–152
  distribution by type of place 160–1
  growth 151–153
  implications of spatial concentration
    164–8
  location within towns 164
  regional distribution 154–60, 162–3
Europe 2

family 2–3

family size 19
female participation rates 98
finance, see local government finance
fiscal migration 55
Forest Heath 11

gender 85
General Household Survey 21
gentrification 38
geodemographics 22
Geographical Information Systems
  145–146
Glasgow 11, 34, 99
Glasgow Quality of Life Group 41
Gloucestershire 81
Gordon 11
government, see local government
Grammar Schools 78
Grant Maintained schools 75–82
Grant Maintained Status 75–82
green belts 115, 116

Harrogate 126
headship rates 105, 111
health care planning 20, 25, 136–49
  deprivation payments to GPs 141–3
  management applications 143–8
  national resource allocation 138–41
Hereford and Worcester 9
Herefordshire 81
Hertfordshire 81, 116
home help 129
Hospital Plan 1962 138
hospitals 125
housebuilding
  future distribution of 114–7
  views on needs 101–3
household composition 3, 112
household dissolution 113
household formation 19, 109–110
household numbers 103–14
  change in 103–4, 109
  components of change 105
  location of growth 106–8
  relationship with housebuilding
    108–14
household size 3, 19, 109
house prices 103
housing 165
housing in poor condition 113
housing market 19
  rigidities in 95
housing needs 2
Housing Needs surveys 102
Humberside 23, 82, 142

Huntingdonshire 11

illness, see limiting long-term illness
immigrants 100
immigration 151, 153–4
income support 130, 133
independent sector homes 126–9
Indians, see ethnic minority populations
indigenous growth 106, 108
industrial districts 12
inner city areas 12, 27, 46, 74, 76, 139
Inner London Education Authority
    (ILEA) 67
internal migration 2
    and quality of life 33–49
    implications 46–7
    migrants' motivations 35–6
    see also migration, labour migration
international migration 2, 12, 105, see
    also immigration
Inverclyde 11
Irish 151–2
Isle of Wight 13
Italians 151–2

Japan 2
Jarman Index 141–2
Jews 151–2

Kent 81
Kincardine and Deeside 11
kin networks 154
Knowsley 11
Korner Committee on Health Services
    Information 145

labour force size
    change by age group 90–93
    local changes 89–90
    regional trends 86–9
Labour Force Survey 21, 100
labour market 19
labour market accounting 94
labour migration 95
labour supply 2
labour supply changes 86–93
    implications 93–9
Lancashire 81
land use policy 114–8
leaflet distribution 25
Leeds 153
Leicester 16
Leicestershire 15
lifestyles 3–4, 38
lifestyle surveys 147

limiting long-term illness 13, 147–148
Lincolnshire 9, 81
Liverpool 11, 67
living costs 36
local authority residential care 126
local dimension, importance of 4–6
local government fees and charges 52
local government finance 18, 50–63
    budgets 59–62
    fiscal balance 53–5
    resources 51–3
local government services 50–51
local taxes 51–5
London 8, 10, 13, 15, 16, 57, 60, 68, 95,
    96, 108, 113, 116, 139, 142, 153
London Docklands 26, 34
long term care homes, see care homes
Lothian 128

Macclesfield 66, 98
Manchester 65–6, 69, 72, 156
marital composition component 105
matching of household and dwelling
    types 112
Merseyside 9, 142
Middlesbrough 137
mid year estimates 7, 100
migrants' motivations 35
migration 4, 6, 8, 18, 29, 33–49,
    106–107, 113
    see also immigration
    internal migration
    international migration
    labour migration
migration and social change 34
Milton Keynes 11, 27
mismatch in housing market 109
mobility transition 4
mortality and deprivation 137
    spatial variations in 137–8
Muslims 78, 167–8
Muslim Parliament of Great Britain
    167–8

National Assistance Act 1948 126
National Child Development Survey 21
National Curriculum 73
National Health Service and Community
    Care Act 1990 119, 133–5
National Health Service Central Register
    21, 27, 48
NHS resource allocation 138–143
natural change 8–9
need for health care
    assessing need 138–143

factors affecting need 136—8
link to mortality 147—8
need index 56—9
needs, measuring 55—6
changing patterns 56—9
new technologies 37
new towns 12, 116
NIMBY 19, 102, 115—6
non-white population, see ethnic
minority populations
Norfolk 9
North America 2
North-South divide 8—11, 34, 87—93,
94
North West 8, 9, 67, see also regional
trends
Northumberland 82, 128
North Yorkshire 81, 130, 132
Nottingham Health Profile 147
Nottinghamshire 82
nursing homes 121
nutrition issues 147

older workers 97—98, 99
one-person households 3
OPCS Longitudinal Survey 21, 48, 124
owner occupation, entry into 110
Oxfordshire 81

Pakistanis see ethnic minority
populations
parental choice 73—5
pensions 2
peripheral council estates 74
'place marketing' 34, 47
planning policies 114—8
point-in-polygon techniques 145
police 24
poll tax 53, 57
population change
by county 9, 23
by district 11, 24
by district type 9—11
by region 7—9
components 8—9
population density 7
population effect 84, 86—7
population estimates 4
updating 29—30
population projections 4, 22—32, 100
case study 27—31
information needs 25—6, 27—8
methods 26—7, 29
users 5, 23—5
Portsmouth 147

Powys 9
primary schools 65—7
private boarding schools 82
private/state continuum in education
79—82
pupil per teacher ratios 70

quality of life 4, 18, 19, 33—49, 113,
118, 119—20
and migration patterns 43—6
and the elderly 119—20
as determinant of migration 36—41
migrants' views 41—4
views by age and position in the life
cycle 39—41

race 78, see also ethnic minority
populations
Race Relations Acts 151, 165
rate capping 71
regional trends
in elderly in long-term care 121—3
in labour force size 86—9
in population 7—9
replacement housing 113
residential care 19
resorts 11
Resource Allocation Working Party
138—41
retailing 28, 31
retail planning 24
retirement areas 11, 12
retirement migration 124
Revenue Support Grant 55—62
riots 165
Rochdale 154
rural areas 37, 57, 74, 98
rural primary schools 66
Rushdie affair 167

Salford 11, 67
school closure 70—72, 75
school places, assessing future demand
74
school rolls 24, 27, 78
effect of GMS 75—9
responses to falling rolls 70—73
role of demographic factors on 65—70
schools 57, 59, 168
opting out 75—82
Scotland 9, 13, 39, 137, see also regional
trends
secondary schools 67—70
Sefton 128
segregation 164—5

service class 46
Shropshire 9
sixth form colleges 69, 79
skills shortages 95−6
small firms 37
social class 74, 137
social security budget 123
Somerset 125, 126, 128, 132,
South Coast 11, 13
South East 8, 9, 34, 39, 45, 66, 98, 112,
    see also regional trends
South West 8, 9, *see also* regional trends
staff transfers 47
Staffordshire 82
Standard Spending Assessment 55−62
staying-on rate 69−70, 95
Stockport 69, 72, 79
Stratford school, East London 78
Strathclyde 9
suburban areas 12
suburbanisation 161, 164
suburbs 66
Suffolk 81, 129
Sunderland 137
Surrey 23, 116

Technical and Vocational Educational
    Initiative (TVEI) 75, 76
telecommunications 38
teleworking 47
Tower Hamlets 11
'town cramming' 101, 115, 118
Townsend index 142

training 165
Training and Enterprise Councils 100
Travel to Work Areas 89
Tyne and Wear 142
Tyneside 156

unemployment 94−5, 165
urban encroachment 101
urban-rural shift 8−11, 60

Wales 8, 13, *see also* regional trends
welfare inequalities 55
West Indians *see also* ethnic minority
    populations
West Midlands 8, 9, 15, 113, 153, *see
    also* regional trends
West Sussex 81
West Yorkshire 15
Winchester 66
Wokingham 11
women 2−3, 121
women in the labour force 98−9
Worcestershire 81
working age population 84
Working Party on Migration
    in Britain 48

Yorkshire and Humberside 8, *see also*
    regional trends
young people in labour market 95−6

zero population growth 1, 9, 46